Apollo 12

The NASA Mission Reports

Compiled from the NASA archives & Edited
by Robert Godwin

All rights reserved under article two of the Berne Copyright Convention (1971).
We acknowledge the financial support of the Government of Canada through the
Book Publishing Industry Development Program for our publishing activities.
Published by Apogee Books an imprint of Collector's Guide Publishing Inc., Box 62034, Burlington, Ontario, Canada, L7R 4K2
Printed and bound in Canada
Apollo 12 - The NASA Mission Reports
by Robert Godwin
ISBN 1-896522-54-8
©1999 Apogee Books
All photos courtesy of NASA

Introduction

The flight of Apollo 12 was nearly flawless. The all-Navy crew of veteran astronauts Charles "Pete"Conrad and Richard "Dick" Gordon along with rookie Alan Bean became famous for their boyish enthusiasm and good humour. Conrad and Gordon had flown together previously on Gemini 11 and their flight experience served them well on Apollo 12.

When things did go wrong they went wrong with a vengeance. Apollo 12 was twice struck by lightning, just after launch, and the cool nerves of Conrad combined with the savvy of Gordon and Bean avoided an abort and what would almost certainly have been a catastrophe for the United States moon program.

Ignoring the banks of flashing lights and the total lack of navigational data, Conrad literally chose to ride out the storm. If ever there was a moment that brought his character into clear relief, this was it.

Most of you reading this probably never got to meet Pete Conrad, neither did I yet when his motorcycle accident was reported on the evening news it was one of the most disheartening moments of my life. Such was the power of the character of this man. Pete's straight ahead speaking style and his cheeky sense of fun pervades the tapes and films of Apollo 12.

When he flew to the moon, Pete Conrad was able to fly his mission almost flawlessly while still making most of us believe that he was just an ordinary guy out for a drive in the park. While the crew of Apollo 11 had, understandably, been all business during the lunar descent, the dialogue coming back from the cockpit of Intrepid was infectious and hilarious. Pete even seems to have surprised his crewmate Al Bean with his enthusiastic cry of, "Hey, there it is! There it is! Son of a gun right down the middle of the road!", Bean replies, "Outstanding, 42 degrees Pete.......42 DEGREES!" not completely sure whether Pete is even listening to him. Conrad continues with, ".....right there in the center of the crater...look out there! I can't believe it! Amazing! Fantastic!" then a pause and finally an acknowledgement, "42 degrees."

In the heat of an incredibly difficult maneuver Bean dutifully continues to rattle off the data from the flight computer while Conrad continues to interject with, "That's so fantastic I can't believe it!" and "Look at that crater, right where it's supposed to be!"

Even thirty years later this amusing exchange is one of the great legacies of Pete Conrad. He made the whole world feel that he was the same as the rest of us while simultaneously making one of the great pin-point landings of all time.

Once it was time to make the descent down the ladder to the lunar surface most of us would probably have been caught up in the moment and uttered a poignant follow-up to Neil Armstrong's great words. Not Pete Conrad.

After making a hilarious remark about the last step of the ladder being, "...a long one for me." in reference to his slight stature, he steps onto the lunar surface with, "Oooh is that soft and cushy." Later when his co-pilot descends and steps into the dust he makes the comforting noises of the relaxed veteran, "Atta-boy you look great. Welcome aboard."

Sadly the other major faux-pas of the flight of Apollo 12 was the sudden demise of the lunar surface television camera. For the fifty or so minutes that the camera worked we were treated to stunningly clear color pictures of Bean and Conrad unpacking the MESA. Unfortunately during training the crew had been obliged to work without the real hardware and so when it came to showtime the camera was pointed into the blinding glare of the sun reflecting off the LM insulation. Despite Al Bean's creative approach of pounding on the camera with his sample hammer, the camera's inner workings were melted and that was it for the Apollo 12 EVA TV. (During reentry the companion 16mm camera decided to avenge its brethren by pounding Bean on the head and inflicting a gash requiring six stitches.)

The crew of Apollo 12 left us with an invaluable scientific legacy. The first full deployment of the ALSEP, the first dead-on pin-point landing on the moon, the first retrieval of parts from another lander, the first observations of an Earth eclipse of the sun, and the successful hi-resolution photography of the lunar surface. However, for my money, the real legacy of Apollo 12 was the way that Gordon, Bean and particularly Pete Conrad showed the world that a bunch of guys with a wicked sense of humour could perform breath-taking miracles.

Robert Godwin
(Editor)

This book is dedicated to
Charles "Pete" Conrad Jr

Here's 12, Pete.

Apollo 12
The NASA Mission Reports
(from the archives of the National Aeronautics and Space Administration)

CONTENTS

APOLLO 12 PRESS KIT

DEDICATION

This press kit is dedicated to the memory of Science Editor William J. Perkinson of The Baltimore Evening Sun and Reporter/Columnist John B. Wilson of The Minneapolis Tribune, who covered the United States space program from its inception through the Apollo 11 lunar landing mission.

MISSION OPERATION REPORT

FIGURES

TABLES

POST FLIGHT MISSION OPERATION REPORT

FIGURES

TABLES

POSTFLIGHT TECHNICAL DEBRIEFING

APOLLO LAUNCH CONFIGURATION FOR LUNAR LANDING MISSION

Command module

Service module

Spacecraft/lunar module adapter

Lunar module

Launch vehicle

Apollo 12 Press Kit

NATIONAL AERONAUTICS & SPACE ADMINISTRATION

FOR RELEASE: WEDNESDAY P.M.
November 5, 1969
RELEASE No: 69-148

PROJECT: APOLLO 12 (To be launched no earlier than Nov. 14)

<u>APOLLO 12 LAUNCH NOV. 14</u>

Apollo 12, the second United States manned lunar landing mission will be launched November 14 from the John F. Kennedy Space Center, Fla., to continue lunar exploration begun last July by the Apollo 11 crew.

Stay time on the Moon will be approximately 10 hours longer than on the first landing last July and the lunar module crew will leave the spacecraft twice to set up scientific experiments and make geological investigations.

Crewmen are commander Charles Conrad, Jr., command module pilot Richard F. Gordon, Jr., and lunar module pilot Alan L. Bean. All are U.S. Navy Commanders.

Primary objectives of Apollo 12 are:

*Perform selenological inspection, survey and sampling in a lunar mare area;

*Deploy an Apollo Lunar Surface Experiments Package (ALSEP) consistent with a seismic net;

*Develop techniques for a point landing capability;

*Develop man's capability to work in the lunar environment; and

*Obtain photographs of candidate exploration sites.

The Apollo 12 landing site is on the western side of the visible face of the Moon in the Ocean of Storms at 2.94° South latitude, 23.45° West longitude — about 830 nautical miles west of Apollo 11's landing site in the Sea of Tranquillity.

The Apollo 12 landing site is designated Site 7.

Experiments in ALSEP 1 will gather and relay long-term scientific and engineering data to Earth for at least a year on the Moon's physical and environmental properties. Six experiments are contained in the ALSEP: lunar passive seismometer for measuring and relaying meteoroid impacts and moonquakes; magnetometer for measuring the magnetic field at the lunar surface; solar wind device for monitoring the interaction of solar wind particles with the Moon; lunar ionosphere detector for measuring flux, energy and velocity of positive ions in the lunar ionosphere; lunar atmosphere detector for measuring minute changes in the ambient lunar atmosphere density; and lunar dust detector for measuring of dust accretion on ALSEP.

During the second of two Extra Vehicular Activity (EVA) periods Conrad and Bean will perform extensive lunar geological investigations and surveys.

A secondary objective is to retrieve portions of the Surveyor III spacecraft which have been exposed to the lunar environment since the unmanned spacecraft soft-landed on the inner slope of a crater April 20, 1967. Inspection and sample collection at the Surveyor landing site will be done during the second EVA period in which Conrad and Bean will make an extended geology traverse. The opportunity to perform this secondary objective is dependent on the LM landing close enough for a traverse to the Surveyor III site.

After the lunar surface phase is complete, a plane-change maneuver with the service propulsion engine will bring the command module over three candidate Apollo landing sites — Fra Mauro, Descartes and Lalande. Extensive photographic coverage will be obtained of these sites for assessment as potential landing points for later Apollo missions.

The flight profile of Apollo 12 in general follows that flown by Apollo 11 with two exceptions: Apollo 12 will have a higher inclination to the lunar equator and will leave the free-return trajectory at midcourse correction No. 2.

Going to a non free-return or hybrid trajectory permits a daylight launch, translunar injection over the Pacific Ocean stretches out the translunar coast to gain the desired landing site lighting at the time of LM landing, and conserves fuel. Also the non free-return allows the 210-foot Goldstone Calif. tracking antenna to cover the LM descent and landing.

A second launch window for the Apollo 12 opens on Nov. 16 for lunar landing Site 5.

Lunar orbit insertion will be made in two phases — one maneuver into a 60x170 nm elliptical orbit, the second maneuver to a more circular orbit of 54x66 nm.

Lunar surface touchdown should take place at 1:53 a.m. EST Nov. 19, and two periods of extravehicular activity are planned at 5:55 a.m. EST Nov. 19 and 12:29 a.m. EST Nov. 20. The LM ascent stage will lift off at 9:23 a.m. Nov. 20 to rejoin the orbiting command module after nearly 32 hours on the lunar surface.

Apollo 12 will leave lunar orbit at 3:43 p.m. EST Nov. 21 for return to Earth. Splashdown in the mid-Pacific just south of the Equator will be at 3:57 p.m. EST Nov. 24.

Apollo 12 backup crew members are USAF Col. David R. Scott, commander; USAF Maj. Alfred M. Worden, command module pilot; and USAF Lt. Col. James B. Irwin., lunar module pilot.

(END OF GENERAL RELEASE; BACKGROUND INFORMATION FOLLOWS)

APOLLO 12 COUNTDOWN

The official countdown for Apollo 12 will begin at T-28 hours and will continue to T-9 hours at which time a built-in hold is planned prior to the start of launch vehicle propellant loading.

Precount activities begin at T-4 days, 2 hours when the space vehicle will be prepared for the start of the official countdown. During precount, final space vehicle ordnance installation and electrical connections will be accomplished. Spacecraft gaseous oxygen and gaseous helium systems will be serviced, spacecraft batteries will be installed, and LM and CSM mechanical buildup will be completed. The CSM fuel cells will be activated and CSM cryogenics (liquid oxygen liquid hydrogen) will be loaded and pressurized.

Following are some of the major operations in the final count:

T-28 hours	Official countdown starts LM crew stowage and cabin closeout (T-31:30 to T-18:00)
T-27 hours, 30 minutes	Install and connect LV flight batteries (to T-23 hours)
T-22 hours, 30 minutes	Top off of LM super critical helium (to T-20 hours, 30 minutes)
T-19 hours, 30 minutes	LM SHe thermal shield installation (to T-15 hours, 30 minutes)
	CSM crew stowage (T-19 to T-12 hours, 30 minutes)
T-16 hours	LV range safety checks (to T-15 hours)
T-15 hours	Installation of ALSEP FCA (to T-14 hours, 45 minutes)
T-11 hours, 30 minutes	Connect LV safe and arm devices (to 10 hours, 45 minutes)
	CSM pre-ingress operations (to T-8 hours, 45 minutes)
T-10 hours, 15 minutes	Start MSS move to park site
T-9 hours.	Built-in-hold for 9 hours and 22 minutes. At end of hold, pad is cleared for LV propellant loading
T-8 hours, 05 minutes	Launch vehicle propellant loading - Three stages (LOX in first stage, LOX and LH2 in second and third stages) Continues through T-3 hours 38 minutes
T-4 hours, 17 minutes	Flight crew alerted
T-4 hours, 02 minutes	Medical examination
T-3 hours, 32 minutes	Breakfast
T-3 hours, 30 minutes	One-hour hold
T-3 hours, 07 minutes	Depart Manned Spacecraft Operations Building for LC-39 via crew transfer van.
T-2 hours, 55 minutes	Arrive at LC-39
T-2 hours, 40 minutes	Start flight crew ingress
T-2 hours	Mission Control Center - Houston/ spacecraft command checks
T-1 hour, 55 minutes	Abort advisory system checks
T-1 hour, 51 minutes	Space Vehicle Emergency Detection System (EDS) test
T-43 minutes	Retract Apollo access arm to standby position (12 degrees)
T-42 minutes	Arm launch escape system
T-40 minutes	Final launch vehicle range safety checks (to 35 minutes)
T-30 minutes	Launch vehicle power transfer test LM switch over to internal power
T-20 minutes to T-10 minutes	Shutdown LM operational instrumentation
T-15 minutes	Spacecraft to internal power
T-6 minutes	Space vehicle final status checks
T-5 minutes, 30 seconds	Arm destruct system
T-5 minutes	Apollo access arm fully retracted
T-3 minutes, 6 seconds	Firing command (automatic sequence)
T-50 seconds	Launch vehicle transfer to internal power
T-8.9 seconds	Ignition sequence start
T-2 seconds	All engines running
T-0	Liftoff

Note: Some changes in the above countdown are possible as a result of experience gained in the countdown demonstration test which occurs about 10 days before launch.

LAUNCH, MISSION TRAJECTORY AND MANEUVER DESCRIPTION

Information presented here is based upon an on-time Nov. 14 launch and is subject to change prior to the mission or in real-time during the mission to meet changing conditions.

Launch

Saturn V launch vehicle will launch the Apollo 12 spacecraft from Launch Complex 39A, NASA-Kennedy Space Center, Fla., on an azimuth that can vary from 72 to 96 degrees, depending upon the time of day of launch. The azimuth changes with launch time to permit a fuel-optimum injection from Earth parking orbit into a free-return circumlunar trajectory and proper Sun angles at the lunar landing site.

November 14 launch plans call for liftoff at 11:22 a.m. EST on an azimuth of 72 degrees. The vehicle will reach an altitude of 36 nm before first stage cutoff 51 nm downrange. During the 2 minutes 42 seconds of powered flight, the first stage will increase vehicle velocity to 9,059 fps (5,363 knots). First stage thrust will reach a maximum of 9,042,041 pounds before center engine cutoff. After engine shutdown and separation from the second stage, the booster will fall into the Atlantic Ocean about 364 nm downrange from the launch site (30 degrees North latitude and 74 degrees West longitude) approximately 9 minutes 14 seconds after liftoff.

The 1-million-pound thrust second stage (S-II) will carry the space vehicle to an altitude of 102 nm and a distance of 884 nm downrange. At engine burnout, the vehicle will be moving at a velocity of 22,889 fps. The outer J-2 engines will burn 6 minutes 26 seconds during the powered phase, but the center engine will be cut off at 4 minutes 57 seconds after S-II ignition.

At outboard engine cutoff, the S-II will separate and, following a ballistic trajectory, plunge into the Atlantic Ocean about 2,419 nm downrange from the Kennedy Space Center (31 degrees North latitude and 34 degrees West longitude) some 20 minutes 24 seconds after liftoff.

The first burn of the Saturn V third stage (S-IVB) begins about 4 seconds after S-II stage separation. It will last long enough (135 seconds) to insert the space vehicle into a circular Earth parking orbit beginning at about 1,429 nm downrange. Velocity at Earth orbital insertion will be 25,567 fps at 11 minutes 39 seconds ground elapsed time (GET). Inclination will be 33 degrees to the equator.

The crew will have a backup to launch vehicle guidance during powered flight. If the Saturn instrument unit inertial platform fails, the crew can switch guidance to the command module systems for first-stage powered flight automatic control. Second and third stage backup guidance is through manual takeover in which crew hand controller inputs are fed through the command module computer to the Saturn instrument unit.

LAUNCH EVENTS

Time Hrs Min Sec	Event	Altitude Feet	Velocity Ft/Sec	Range Nau. Mi.
00 00 00	First Motion	198	1,340	0
00 01 23	Maximum Dynamic Pressure	44,250	2,702	3
00 02 15	S-IC Center Engine Cutoff	143,972	6,454	25
00 02 42	S-IC Outboard Engines Cutoff	222,090	9,058	51
00 02 43	S-IC S-II Separation	224,529	9,089	52
00 02 44	S-II Ignition	229,619	9,075	52
00 03 13	S-II Aft Interstage Jettison	310,032	9,485	88
00 03 18	LET Jettison	324,151	9,588	95
00 07 41	S-II Center Engine Cutoff	608,352	17,582	603
00 09 11	S-II Outboard Engines Cutoff	623,592	22,888	884
00 09 12	S-II/S-IVB Separation	623,782	22,897	887
00 09 15	S-IVB Ignition	624,326	22,898	898
00 11 30	S-IVB First Cutoff	627,958	25,562	1,390
00 11 40	Parking Orbit Insertion	627,981	25,567	1,429
02 47 20	S-IVB Reignition	644,318	25,559	5,716
02 53 05	S-IVB Second Cutoff	1,162,582	35,426	4,165
02 53 15	Translunar Injection	1,214,454	35,394	4,112

APOLLO 12 MISSION EVENTS

Event	GET Hrs : min.	Date/EST	Vel. Change feet/sec	Purpose and Resultant Orbit
Earth orbit insertion	00:11	14 11:34 a.m.	25,567	Insertion into 103 nm circular Earth parking orbit.
Translunar injection (S-IVB engine ignition)	02:47	14 2:09 p.m.	9,859	Injection into free-return translunar trajectory with 1,850 nm pericynthion
CSM separation, docking	03:28	14 2:50 p.m.		Hard-mating of CSM and LM
Ejection from SLA	04:13	14 3:35 p.m.	1	Separates CSM-LM from S-IVB-SLA
S-IVB Evasive maneuver	04:25	14 3:47 p.m.	10	Provides separation prior to S-IVB propellant dump and "slingshot" maneuver
Midcourse correction #1	TLI+9 hrs	14 11:09 p.m.	*0	*These midcourse corrections have a nominal velocity change
Midcourse correction #2 (Hybrid transfer)	30:53	15 6:15 p.m.	64	of 0 fps, but will be calculated in real time to correct TLI dispersions. MCC-2 is an
Midcourse correction #3	LOI-22 hrs	16 00:47 a.m.	*0	SPS maneuver (64 fps) to lower pericynthion to 60 nm; trajectory then becomes non-free return.
Midcourse correction #4	LOI-5 hrs	17 5:47 p.m.	*0	
Lunar orbit insertion #1	83:25	17 10:47 p.m.	-2890	Inserts Apollo 12 into 60x170nm elliptical lunar orbit
Lunar orbit insertion #2	87:44	18 3:06 a.m.	-169	Changes lunar parking orbit to 54x66 nm
CSM-LM undocking	107:58	18 11:20 p.m.		Establishes equiperiod orbit for 2.2 nm separation at DOI maneuver
Separation (SM RCS)	108:28	18 11:49 p.m.	2.5	
Descent orbit insertion	109:23	19 00:45 a.m.	-72	Lowers LM pericynthion to 8 nm (8x60)
LM powered descent	110:20	19 1:42 a.m.	-6779	Three-phase maneuver to initiation (DPS)brake LM out of transfer orbit, vertical descent and touchdown on lunar surface
LM touchdown on lunar surface	110:31	19 1:53 a.m.		Lunar exploration, deploy ALSEP, lunar surface geological sample collection, photography, possible Surveyor III investigation
Depressurization for 1st lunar surface EVA	114:33	19 5:55 a.m.		
CDR steps to surface	114:47	19 6:09 a.m.		
CDR collects contingency samples		19 6:16 a.m.		
LMP steps to surface	115:14	19 6:36 a.m.		
CDR unstows and erects S-Band antenna	115:20	19 6:42 a.m.		
LMP mounts TV camera on tripod	115:26	19 6:48 a.m.		
LMP deploys solar wind experiment	115:33	19 6:55a.m.		
CDR and LMP begin unstowing and deployment of ALSEP	115:46	19 7:08a.m.		
CDR and LMP return to LM collecting samples and retrieving TV camera on route	117:00	19 8:22 a.m.		
CDR and LMP arrive back at LM, stow equipment and samples	117:17	19 8:39 a.m.		
LMP reenters LM	117:33	19 8:55 a.m.		

APOLLO SPACECRAFT/LM ADAPTER

PANEL SEPARATION BY EXPLOSIVE CHARGES (MDF)

Event	GET	Day	Time	ΔV	Description
CDR reenters LM	117:53	19	9:15 a.m.		
LM hatch closed and repressurize	117:58	19	9:20 a.m.		
Depressurization for 2nd lunar surface EVA	133:07	20	00:29 a.m.		
CDR steps to surface	133:13	20	00:35 a.m.		
LMP steps to surface	133:20	20	00:42 a.m.		
Begin field geology traverse & collect core tube & gas analysis sample	133:30	20	00:52 a.m.		
Walk to Surveyor Site, observe, photograph & retrieve parts	135:00	20	2:22 a.m.		
Complete geology traverse	135:30	20	2:52 a.m.		
Return to LM area retrieve solar wind experiment; stow surface samples	135:45	20	3:07 a.m.		
LMP enters LM	135:59	20	3:21 a.m.		
CDR transfers samples LMP assists	136:09	20	3:31 a.m.		
CDR enters LM and closes hatch	136:24	20	3:46 a.m.		
Cabin repressurization	136:30	20	3:52 a.m.		
LM ascent	142:01	20	9:23 a.m.	6049	Boosts ascent stage into 9x45 lunar orbit for rendezvous with CSM
Insertion	142:08	20	9:30 a.m.		
LM RCS concentric sequence initiation (CSI) burn	142:58	20	10:20 a.m.	50	Raises LM perilune to 44.7 nm, adjusts orbital shape for rendezvous sequence (47x45)
LM RCS constant delta burn	143:56	20	11:18 a.m.	4.4	Radially downward burn adjusts height (CDH) LM orbit to constant 15 nm below CSM
LM RCS terminal phase initiation (TPI) burn	144:36	20	11:58 a.m.	24.8	LM thrusts along line of sight toward CSM, midcourse and braking maneuvers as necessary
Rendezvous (TPF)	145:21	20	12:43 p.m.	31.7	*Completes rendezvous sequence (59.5x59.0)
Docking	145:40	20	1:02 p.m.		Commander and LM pilot transfer back to CSM
LM jettison separation (SM RCS)	147:21	20	2:43 p.m.	1.5	Prevents recontact of CSM with LM ascent stage during remainder of lunar orbit
LM ascent stage deorbit	149:28	20	4:50 p.m.	-200	Seismometer calibration(APS)
LM ascent stage impact	149:56	20	5:18 p.m.		Impact at about 5500 fps, at 4° angle 5 nm from ALSEP
Plane change for photos	159:02	21	2:24 a.m.		
Transearth injection	172:21	21	3:43 p.m.	3113	Inject CSM into transearth(TEI) SPS trajectory
Midcourse correction #5	187:21	22	6:43 a.m.	0	Transearth midcourse corrections will be computed in real time for entry corridor
Midcourse correction #6	EI-22 hrs	23	5:49 p.m.	0	control and recovery area
Midcourse correction #7	EI-3 hrs	24	12:49 p.m.	0	weather avoidance.
CM/SM separation	244:11	24	3:34 p.m.		Command module oriented for entry
Entry interface	244:26	24	3:43 p.m.		Command module enters (400,000 feet) Earth's sensible atmosphere at 36,129 fps
Splashdown	244:35	24	3:57 p.m.		Landing 1250 nm downrange from entry, 16° south latitude by 165° west longitude. (Local time — 9:57 a.m.) Sunrise + 5 hours.

APOLLO CSM & LM COMPARISON

CSM

LM

Earth Parking Orbit (EPO)

Apollo 12 will remain in Earth parking orbit for one and one-half revolutions. The final "go" for the TLI burn will be given to the crew through the Carnarvon, Australia, Manned Space Flight Network station.

Translunar Injection (TLI)

Midway through the second revolution in Earth parking orbit, the S-IVB third-stage engine will restart at 2:47 GET over the mid-Pacific ocean near the equator to inject Apollo 12 toward the Moon. The velocity will increase from 25,559 fps to 35,426 fps at TLI cutoff to return circumlunar trajectory from which midcourse corrections could be made with the SM RCS thrusters. Entry from a free-return trajectory would be at 6:52 a.m. EST Nov. 21 at 2.4 degrees north latitude by 92 west longitude after a flight time of 163 hrs 30 mins. A free-return trajectory from TLI would have an 1850 nm pericynthion at the Moon.

Transposition, Docking and Ejection (TD&E)

After the TLI burn, the Apollo 12 crew will separate the command/service module from the spacecraft lunar module adapter (SLA), thrust out away from the S-IVB, turn around and move back in for docking with the lunar module. Docking should take place at about three hours and 28 minutes GET. After the crew confirms all docking latches solidly engaged, they will connect the CSM to LM umbilicals and pressurize the LM with oxygen from the command module surge tank. At about 4:13 GET, the spacecraft will be ejected from the spacecraft LM adapter by spring devices at the four LM landing gear "knee" attach points. The ejection springs will impart about one fps velocity to the spacecraft. A 10 fps S-IVB attitude thruster evasive maneuver in plane at 4:25 GET will separate the spacecraft to a safe distance for the S-IVB "slingshot" maneuver. The "slingshot" is achieved by burning S-IVB auxiliary propulsion system subsequent to the dumping of residual liquid oxygen through the J-2 engine bell to propel the stage into a trajectory passing the Moon's trailing edge and into solar orbit.

Translunar Coast

Up to four midcourse correction burns are planned during the translunar coast phase, depending upon the accuracy of the trajectory resulting from the TLI maneuver. If required, the midcourse correction burns are planned at TLI+9 hours, TLI+31 hours, lunar orbit insertion (LOI) - 22 hours and LOI - 5 hours. The MCC-2 at TLI+31 hrs is a 64 fps SPS hybrid transfer maneuver which lowers pericynthion from 1852 nm to 60 nm and places Apollo 12 on a non free-return trajectory. Return to the free-return trajectory is always within the capability of the spacecraft service propulsion or descent propulsion systems.

During coast periods between midcourse corrections, the spacecraft will be in the passive thermal control (PTC) or "barbecue" mode in which the spacecraft will rotate slowly about its roll axis to stabilize spacecraft thermal response to the continuous solar exposure.

Lunar Orbit Insertion (LOI)

The first of two lunar orbit insertion burns will be made at 83:25 GET at an altitude of about 85 nm above the Moon. LOI-1 will have a nominal retrograde velocity change of 2890 fps and will insert Apollo 12 into a 60x169 nm elliptical lunar orbit. LOI-2 two orbits later at 87:44 GET will adjust the orbit to a 54x66 nm orbit, which because of perturbations of the lunar gravitational potential, should become circular at 60 nm at the time of rendezvous with the LM. The burn will be 169 fps retrograde. Both LOI maneuvers will be with the SPS engine near pericynthion when the spacecraft is behind the Moon, out of contact with MSFN stations. After LOI-2 (circularization), the commander and lunar module pilot will enter the lunar module for a brief checkout and return to the command module.

Lunar Module Descent - Lunar Landing

The lunar module will be manned and checked out for undocking and subsequent landing on the lunar surface at Apollo site 7. Undocking will take place at 107:58 GET prior to the MSFN acquisition of signal. A radially downward service module RCS burn of 2.5 fps will place the CSM on an equiperiod orbit with a maximum separation of 2.2 nm one half revolution after the separation maneuver. At this point, on lunar farside, the descent orbit insertion burn (DOI) will be made with the lunar module descent engine firing retrograde 72 fps at 109:23 GET. The burn will start at 10 percent throttle for 15 seconds and the remainder at 40 percent throttle.

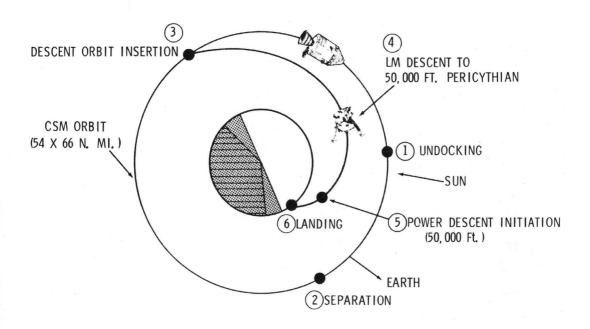

LUNAR DESCENT EVENTS

The DOI maneuver lowers LM pericynthion to 50,000 feet at a point about 15 degrees uprange of landing site 7.

A three-phase powered descent initiation (PDI) maneuver begins at pericynthion at 110:20 GET using the LM descent engine to brake the vehicle out of the descent transfer orbit. The guidance-controlled PDI maneuver starts about 260 nm prior to touchdown, and is in retrograde attitude to reduce velocity to essentially zero at the time vertical descent begins. Spacecraft attitude will be windows up from powered descent to landing so that the LM landing radar data can be integrated continually by the LM guidance computer and better communications can be maintained. The braking phase ends at about 7,000 feet above the surface and the spacecraft is rotated to an upright windows-forward attitude. The start of the approach phase is called high gate, and the start of the landing phase at 500 feet is called low gate.

Both the approach phase and landing phase allow pilot takeover from guidance control as well as visual evaluation of the landing site. The final vertical descent to touchdown begins at about 100 feet when all forward velocity is nulled out. Vertical descent rate will be three fps. Present plans call for the crew to take over manual control at approximately 500 feet. Touchdown will take place at 110:31 GET.

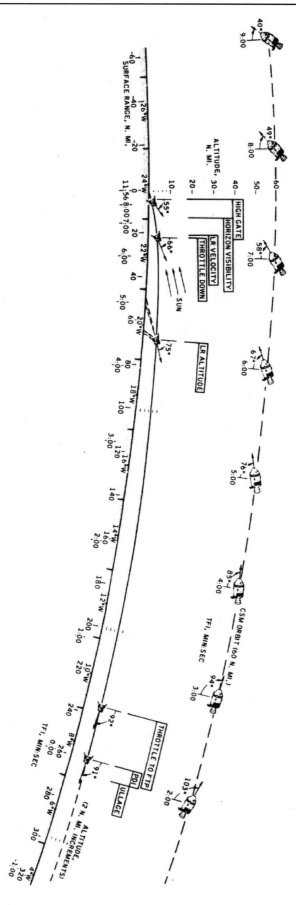

POWERED DESCENT PROFILE

Powered Descent Events

Summary

Event	TPI Min:Sec	Velocity FPS	Altitude Rate	Height Feet
Powered Descent Initiation	0:00	5560	-4	49,235
Throttle to Maximum Thrust	0:26	5529	-3	49,158
Landing Radar Altitude Update	3:52	2959	-94	39,186
Throttle Recovery	6:24	1522	-106	24,592
Landing Radar Velocity Update	6:40	1301	-136	22,945
Horizon Visibility	7:04	1162	-148	20,130
High Gate	8:28	505	-137	7,335
Low Gate	10:22	35 (47)*	-14	524
Touchdown (Probe Contact)	11:56	-15 (0)*	-3	12

* (Horizontal Velocity Relative to Surface)

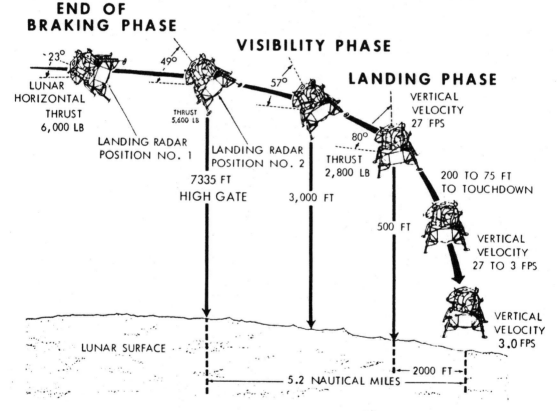

END OF BRAKING PHASE

VISIBILITY PHASE

LANDING PHASE

-23°

LUNAR HORIZONTAL THRUST 6,000 LB

49°

THRUST 5,600 LB

57°

80°

VERTICAL VELOCITY 27 FPS

LANDING RADAR POSITION NO. 1

LANDING RADAR POSITION NO. 2

THRUST 2,800 LB

200 TO 75 FT TO TOUCHDOWN

7335 FT HIGH GATE

3,000 FT

500 FT

VERTICAL VELOCITY 27 TO 3 FPS

VERTICAL VELOCITY 3.0 FPS

LUNAR SURFACE

2000 FT

5.2 NAUTICAL MILES

NOMINAL DESCENT TRAJECTORY
FROM HIGH GATE TO TOUCHDOWN

Lunar Surface Exploration

The manned lunar exploration begun by Apollo 11 will be broadened in the two periods of lunar surface extravehicular activity planned for Apollo 12. In addition to gathering more data on the lunar surface environment and bringing back geological samples from a second landing site, the Apollo 12 crew will deploy a series of experiments which will relay back to Earth long-term scientific measurements of the Moon's physical and environmental properties.

The experiment series, called the Apollo Lunar Surface Experiment Package (ALSEP 1), will remain on the surface to transmit scientific and engineering data to the Manned Space Flight Network for at least a year.

ALSEP 1 comprises equipment for the lunar passive seismic, lunar surface tri-axis magnetometer, solar wind spectrometer, and lunar ionosphere and atmosphere detectors. Additionally, three non-ALSEP experiments — solar wind composition, field geology and Apollo lunar close-up camera photography — will be deployed.

Experiments are aimed toward determining the structure and state of the lunar interior, the composition and structure of the lunar surface and the processes which modified the surface, and the evolutionary sequence leading to the Moon's present characteristics.

Apollo 12's stay on the lunar surface is planned not to exceed 31.5 hours, during which time Conrad and Bean will twice leave the LM to deploy the ALSEP, gather geologic samples and conduct experiments. The crew's operating radius will be limited by the range provided by the oxygen purge system (OPS) mounted atop each man's portable life support system (PLSS) backpack. The OPS supplies one-half hour of emergency breathing oxygen and suit pressure.

Among the tasks assigned Conrad and Bean for the two EVA periods are:

*Collecting a contingency sample near the LM of about two pounds of lunar material.

*Evaluating crew ability to perform useful work in the lunar environment, such as lifting and maneuvering large packages, unstowing and erecting the S-Band antenna. Also, the crew will assess their ability to move about on the lunar terrain and their ability to meet timelines.

*Inspecting the LM exterior for effects of landing impact and lunar surface erosion from descent engine plume.

*Gathering about 30 to 60 pounds of representative lunar surface material, including a core sample, individual rock samples and fine-grained fragments. The crew will photograph thoroughly the areas from which samples are taken.

*Making observations and gathering data on the mechanical properties and terrain characteristics of the lunar surface and conducting other lunar field geological surveys.

An added bonus of Apollo 12's EVA may be the retrieval of portions of the Surveyor III spacecraft which have been exposed to the lunar environment since the spacecraft soft-landed April 20, 1967 at 3.3° south latitude by 23° west longitude. The Surveyor rests approximately 150 feet down a 14 degree slope of a crater that is 50 feet deep and about 650 feet wide.

Surveyor III's television camera sent back 6,315 pictures of the terrain features in the crater in which the spacecraft landed. The 17-pound TV camera, with its variety of electronic and mechanical components will be one of the major items Conrad and Bean hope to retrieve. A modified pair of bolt cutters and a stowage bag will be carried for removing and retrieving Surveyor components.

Before Conrad and Bean begin snipping off portions of the Surveyor spacecraft, they will take color photographs of areas covered by the Surveyor TV camera to show the comparative change, if any, of the surface features.

The trek to the Surveyor landing site will depend upon the accuracy of the Apollo 12 lunar module landing. If the LM lands within one kilometer (3,300 feet) of Surveyor, the landing crew will make a traverse to Surveyor during the second EVA period.

Ascent, Lunar Orbit Rendezvous

Following the 31.5 hour lunar stay the LM ascent stage will lift off the lunar surface to begin the rendezvous sequence with the orbiting CSM. Ignition of the LM ascent engine will be at 142:01 for a 7 min 11 sec burn with a total velocity of 5535 fps. Powered ascent is in two phases: vertical ascent for terrain clearance and the orbital insertion phase. Pitchover along the desired launch azimuth begins as the vertical ascent rate reaches 50 fps about 10 seconds after liftoff at about 272 feet in altitude. Insertion into a 9 x 45-nm lunar orbit will take place about 166 nm west of the landing site.

Following LM insertion into lunar orbit, the LM crew will compute onboard the major maneuvers for rendezvous with the CSM which is about 260 nm ahead of the LM at this point. All maneuvers in the sequence will be made with the LM RCS thrusters. The premission rendezvous sequence maneuvers, times and velocities which likely will differ slightly in real time, are as follows:

Concentric sequence initiate (CSI): At first LM apolune after insertion, 142:58 GET, 50 fps posigrade following some 20 minutes of LM rendezvous radar tracking and CSM sextant/VHF ranging navigation. CSI will be targeted to place the LM in an orbit 15 nm below the CSM at the time of the later constant delta height (CDH) maneuver.

The CSI burn may also initiate corrections for any out-of-plane dispersions resulting from insertion azimuth errors. Resulting LM orbit after CSI will be 46.7 x 44.5 nm and will have a catch-up rate to the CSM of .07 degrees per minute.

Another plane correction is possible about 30 minutes after CSI at the nodal crossing of the CSM and LM orbits to place both vehicles at a common node at the time of the CDH maneuver at 143:56 GET.

Terminal phase initiation (TPI). This maneuver occurs at 144:36 and adds 24 fps along the line of sight toward the CSM when the elevation angle to the CSM reaches 26.6 degrees. The LM orbit becomes 61.9 x 43.8 nm and the catch-up rate to the CSM decreases to .033 degrees per second, or a closing rate of 13.3 fps.

Midcourse correction maneuvers will be made if needed, followed by four braking maneuvers. Docking nominally will take place at 145:40 GET to end three and one-half hours of the rendezvous sequence.

The LM ascent stage will be jettisoned at 147:21 GET and CSM RCS 1.5 fps maneuver will provide separation.

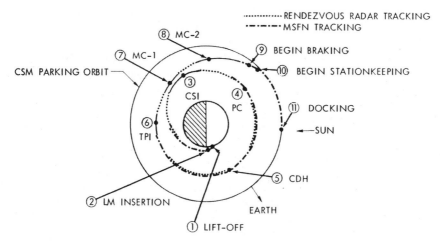

ASCENT THROUGH DOCKING

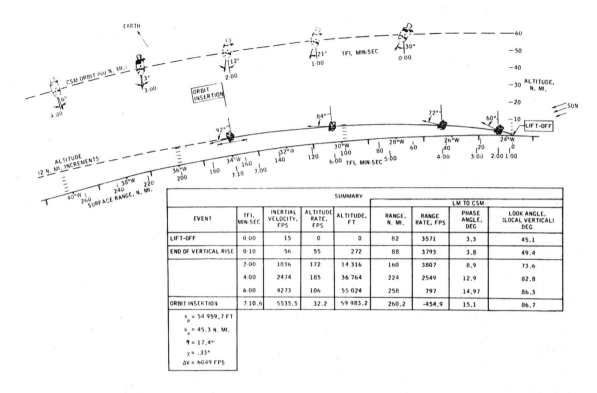

EVENT	TFI, MIN:SEC	INERTIAL VELOCITY, FPS	ALTITUDE RATE, FPS	ALTITUDE, FT	RANGE, N. MI.	RANGE RATE, FPS	PHASE ANGLE, DEG	LOOK ANGLE, (LOCAL VERTICAL) DEG
						LM TO CSM		
LIFT-OFF	0:00	15	0	0	82	3571	3.3	45.1
END OF VERTICAL RISE	0:10	56	55	272	88	3793	3.8	49.4
	2:00	1036	172	14 316	160	3807	8.9	73.6
	4:00	2474	185	36 764	224	2549	12.9	82.8
	6:00	4273	106	55 024	258	797	14.97	86.3
ORBIT INSERTION	7:10.6	5535.5	32.2	59 983.2	260.2	-454.9	15.1	86.7

SUMMARY

$h_p = 54\ 959.7\ FT$

$h_a = 45.3\ N.\ MI.$

$\eta = 17.4°$

$\gamma = .33°$

$\Delta V = 6049\ FPS$

ORBIT INSERTION PHASE

Ascent Stage Deorbit

Prior to transferring to the command module, the LM crew will set up the LM guidance system to maintain the ascent stage in an inertial attitude. At about 149:28 GET the LM ascent engine will ignite on ground command for 200 fps retrograde burn targeted for ascent stage impact at 149:56 about five nm south of Site 7. The burn will have a small out-of-plane north component to drive the stage back toward the ground track of the original landing site. The ascent stage will impact at about 5500 fps at an angle of four degrees relative to the local horizontal. Impacting an object with a known velocity and mass near the landing site will provide experimenters with an event for calibrating readouts from the ALSEP seismometer left behind at Site 7. The ascent stage deorbit also serves to remove debris from lunar orbit.

A plane change maneuver at 159:02 GET will place the CSM on an orbital track passing directly over the craters Descartes and Fra Mauro two revolutions later. The maneuver will be a 360-fps SPS burn out of plane for a plane change of 3.2 degrees.

The Apollo 12 crew will obtain extensive photographic coverage of these lunar surface features.

Transearth Injection (TEI)

The nominal transearth injection burn will be at 172:21 GET following 89 hours in lunar orbit. TEI will take place on the lunar farside, will be a 3113 fps posigrade SPS burn of 2 min 17 sec duration and will produce an entry velocity of 36,129 fps after a 72 hr transearth flight time.

Transearth Coast

Three corridor-control transearth midcourse correction burns will be made if needed: MCC-5 at TEI+15 hrs, MCC-6 at entry interface (EI) -22 hrs and MCC-7 at EI -3 hrs.

Entry, Landing

Apollo 12 will encounter the Earth's atmosphere (400,000 feet) at 244:26 GET at a velocity of 36,129 fps and will land approximately 1250 nm downrange from the entry-interface point using the spacecraft's lifting characteristics to reach the landing point. Splashdown will be at 244:35 at 16 degrees south latitude by 165 degrees west longitude.

Recovery Operations

The prime recovery line for Apollo 12 is the mid-Pacific along the 175th west longitude above 15 degrees north latitude, and jogging to 165 degrees west longitude below the equator. The aircraft carrier USS Hornet, Apollo 12 prime recovery ship, will be stationed near the end-of-mission aiming point prior to entry.

Splashdown for a full-duration lunar landing mission launched on time November 14 will be at 16 degrees South by 165 degrees west.

Launch abort landing areas extend downrange 3,400 nautical miles from Kennedy Space Center, fanwise 50 nm above and below the limits of the variable launch azimuth (72-96 degrees). Ships on station in the launch abort area will be the destroyer USS Hawkins, and the insertion tracking ship USNS Vanguard. The landing platform helicopter, USS Austin will be stationed further south in the Atlantic to support the possible aborts during translunar coast.

In addition to the primary recovery ship located on the mid-Pacific recovery line and surface vessels on the Atlantic Ocean recovery line (along 30th west meridian north) and in the launch abort, thirteen HC-130 aircraft will be on standby at eight staging bases around the Earth: Guam; Hawaii; American Samoa; Bermuda; Lajes, Azores; Ascension Island; Mauritius and the Panama Canal Zone.

Apollo 12 recovery operations will be directed from the Recovery Operations Control Room in the Mission Control Center, supported by the Atlantic Recovery Control Centers Norfolk, Va. and the Pacific Recovery Control Center, Kunia, Hawaii.

After splashdown, the Apollo 12 crew will don biological isolation garments passed to them through the spacecraft hatch by a recovery swimmer. The crew will be carried by helicopter to Hornet where they will enter a Mobile Quarantine Facility (MQF) about 90 minutes after landing.

APOLLO 12 ONBOARD TELEVISION

Two color television cameras are planned to be carried aboard Apollo 12 - one in the command module and one in the lunar module descent stage to transmit a real-time picture of the two periods of lunar surface extravehicular activity. A black and white TV camera may be substituted for the LM color camera.

Both cameras have been refurbished and modified from previous missions; the Apollo 10 command module camera will be stowed in the LM for lunar surface TV, and the Apollo 11 command module camera will be used in the Apollo 12 command module.

The color TV cameras weigh 12 pounds and are fitted with zoom lens for wide-angle or close-up fields of view. The command module camera is fitted with a three-inch monitor which can be detached and placed at a convenient location in the CM. The LM camera will be aimed and focused by the LM crew during EVA with the help of Mission Control.

Built by Westinghouse Electric Corp. Aerospace Division, Baltimore Md. the color cameras output a standard 525-line, 30 frame-per-second signal in color by use of rotating color wheels. The black and white signals carried on the S-Band down-link will be converted to color at the Mission Control Center.

Modifications to the LM color camera include painting it white for thermal control substituting coated metal gears for plastic gears in the color wheel drive mechanism, provision for internal heat conduction paths to the camera outer shell for radiation, and use of a special bearing lubricant.

The lunar module black and white television camera weighs 7.25 pounds and draws 6.5 watts of 24-32 volts DC power. Scan rate is 10 frames-per-second at 320 lines-per-frame. The camera body is 10.6 inches long, 6.5 inches wide and 3.4 inches deep. The bayonet lens mount permits lens changes by a crewman in a pressurized suit. Two lenses, a wide-angle lens for close-ups and large areas, and a lunar day lens for viewing lunar surface features and activities in the near field of view with sunlight illumination, will be provided for the lunar TV camera.

APOLLO 12 TELEVISION DEPLOYMENT

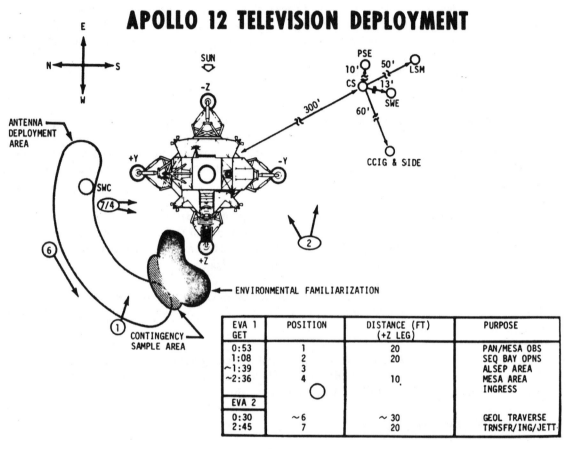

EVA 1 GET	POSITION	DISTANCE (FT) (+Z LEG)	PURPOSE
0:53	1	20	PAN/MESA OBS
1:08	2	20	SEQ BAY OPNS
~1:39	3		ALSEP AREA
~2:36	4	10	MESA AREA INGRESS
EVA 2			
0:30	~6	~30	GEOL TRAVERSE
2:45	7	20	TRNSFR/ING/JETT

APOLLO 12 TV SCHEDULE

DAY	DATE	EST	GET	COVERAGE	DURATION
FRIDAY	NOV. 14	14:42	03:25	TRANSPOSITION/DOCKING	1 + 05
SATURDAY	NOV. 15	17:47	30:25	HYBRID TRAJ./SPACECRAFT INTERIOR	0 + 35
MONDAY	NOV. 17	02:52	63:30	EARTH, IVT, S/C INTERIOR	0 + 50
		20:52	81:30	PRE LOI-1, LUNAR SURFACE	0 + 20
		23:22	84:00	LUNAR SURFACE	0 + 30
TUESDAY	NOV. 18	23:12	107:50	UNDOCKING/FORMATION FLYING	0 + 20
WEDNESDAY	NOV. 19	06:02	114:40	LUNAR SURFACE EVA	3 + 30
THURSDAY	NOV. 20	00:42	133:20	EVA - 2, EQUIPMENT JETTISON	4 + 55
		12:37	145:15	DOCKING	0 + 30
FRIDAY	NOV. 21	16:17	*172:55	POST - TEI/LUNAR SURFACE	0 + 20
SUNDAY	NOV. 23	06:52	223:15	MOON - EARTH - S/C INTERIOR	0 + 30

*Plans call for this TV event to be recorded for later playback.

APOLLO 12 SCIENTIFIC EXPERIMENTS

The Moon's surface is bombarded by the solar wind which consists of charged particles, mostly protons and electrons, emanating from the Sun. There also exists an interplanetary magnetic field which is carried from the Sun by the solar wind.

The Earth has its own magnetic field which protects it from the direct stream of solar wind charged particles and the solar magnetic flux. The Moon, however, has only a small or negligible magnetic field of its own. As a result the Moon is subject to forces of the solar winds, the solar magnetic field and, during certain times, to the Earth's magnetic field.

Several experiments in the Apollo Lunar Surface Experiments Package (ALSEP) will measure these influences both in sunlight and in darkness on the Moon and, as the Moon passes through the region of the Earth's magnetic field (magnetosphere). Other experiments will obtain information on the physical properties of the Moon's surface and its interior.

The Lunar Surface Magnetometer (LSM)

The scientific objective of the magnetometer experiment is to measure the magnetic field at the lunar surface. Charged particles and the magnetic field of the solar wind impact directly on the lunar surface. Some of the solar wind particles are absorbed by the surface layer of the Moon. Others may be deflected around the Moon. The electrical properties of the material making up the Moon determine what happens to the magnetic field when it hits the Moon. If the Moon is a perfect insulator the magnetic field will pass through the Moon undisturbed, if there is material present which acts as a conductor, electric currents will flow in the Moon.

Two possible models are shown in the next drawing. The electric current carried by the solar wind goes through the Moon and "closes" in the space surrounding the Moon, (figure a). This current (E) generates a magnetic field (M) as shown. The magnetic field carried in the solar wind will set up a system of electric currents in the Moon or along the surface. These currents will generate another magnetic field which tries to counteract the solar wind field, (figure b). This results in a change in the total magnetic field measured at the lunar surface.

LUNAR ENVIRONMENT

MAGNETOMETER   SOLAR WIND SPECTROMETER   PRESSURE GAUGE / ION DETECTOR

MOON - SOLAR WIND INTERACTION

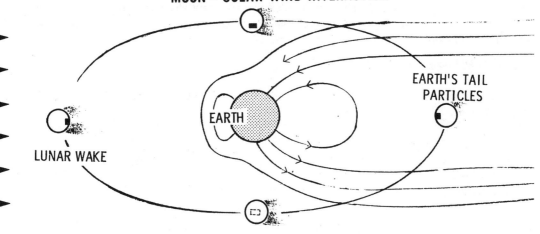

LUNAR WAKE

EARTH

EARTH'S TAIL PARTICLES

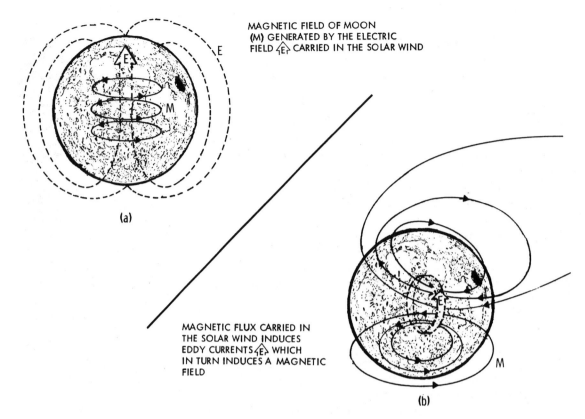

MAGNETIC FIELD OF MOON
(M) GENERATED BY THE ELECTRIC
FIELD $\langle E \rangle$ CARRIED IN THE SOLAR WIND

(a)

MAGNETIC FLUX CARRIED IN
THE SOLAR WIND INDUCES
EDDY CURRENTS $\langle E \rangle$ WHICH
IN TURN INDUCES A MAGNETIC
FIELD

(b)

The magnitude of this difference can be determined by independently measuring the magnetic field in the undisturbed solar wind nearby, yet away from the Moon's surface. It is planned to obtain this data from Explorer 35, the unmanned spacecraft now in lunar orbit. The value of the magnetic field change at the Moon's surface can then be used to deduce information on the electrical properties of the Moon. This, in turn, can be used to better understand the internal temperature of the Moon and contribute to better understanding of the origin and history of the Moon.

The design of the tri-axis flux-gate magnetometer and analysis of experiment data are the responsibility of Dr. Charles P. Sonett NASA/Ames Research Center; Dr. Jerry Modisette - NASA/Manned Spacecraft Center; and Dr. Palmer Dyal - NASA/Ames Research Center.

The magnetometer consists of three magnetic sensors aligned in three orthogonal sensing axes, each located at the end of a fiberglass support arm extending from a central structure. This structure houses both the experiment electronics and the electromechanical gimbal/flip unit which allows the sensor to be pointed in any direction for site survey and calibration modes. The astronaut aligns the magnetometer experiment to within ±3° East-West using a shadow graph on the central structure, and to within ±3° of the vertical using a bubble level mounted on the Y sensor boom arm.

Size, weight and power are as follows:

Size (inches) deployed	40 high with 60 between sensor heads
Weight (pounds)	17.5

Peak Power Requirements (watts)

Site Survey Mode	11.5
Scientific Mode	6.2
	12.3 (night)
Calibration Mode	10.8

The magnetometer experiment operates in three modes:

<u>Site Survey Mode</u> — An initial site survey is performed in each of the three sensing modes for the purpose of locating and identifying any magnetic influences permanently inherent in the deployment site so that they will not affect the interpretation of the ME sensing of magnetic flux at the lunar surface. Although no measurable lunar magnetic field has been detected to date, the possibility of localized magnetism remains; thus this precaution must be taken.

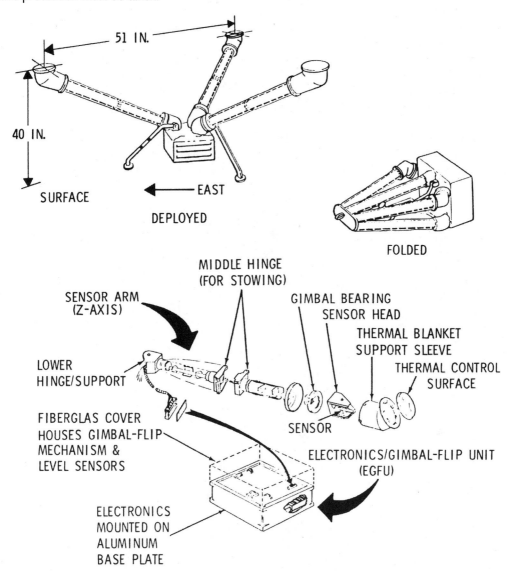

LUNAR SURFACE MAGNETOMETER

<u>Scientific Mode</u> — This is the normal operating mode wherein the strength and direction of the lunar magnetic field are measured continuously. The three magnetic sensors provide signal outputs proportional to the incidence of magnetic field components parallel to their respective axes. Each sensor will record the intensity three times per second which is faster than the magnetic field is expected to change. All sensors have the capability to sense over any one of three dynamic ranges with a resolution of 0.2 gamma.*

-100 to +100 gamma
-200 to +200 gamma
-400 to +400 gamma

*Gamma is a unit of intensity of a magnetic field. The Earth at the equator, for example, is 35,000 gamma. The interplanetary magnetic field from the Sun has been recorded at 5 to 10 gamma.

<u>Calibration Mode</u> — This is performed automatically at 12 hour intervals to determine the absolute accuracy of the magnetometer sensors and to correct any drift from their laboratory calibration.

<u>Lunar Ionosphere Detector (LID)</u> — The scientific objective of the Lunar Ionosphere Detector is to study the charged particles in the lunar atmosphere. In conjunction with the Lunar Atmosphere Detector both charged and neutral particles will be measured by Apollo 12 ALSEP. Although the amount of material detected is expected to be very small, knowledge of the lunar ionosphere density and composition will contribute to the understanding of the Moon's chemistry, radioactivity, and volcanic activity and to the chemical composition of the solar wind. Elements of the solar wind are expected to be the major ionosphere component, but observation from Earth of lunar "hot spots" suggest gas is being released from the Moon. The impact of meteorites on the lunar surface will vaporize both the meteorite and lunar surface material. While all these factors contribute material to the lunar ionosphere and atmosphere, forces are at work contributing to their escape, i.e., the low gravity of the Moon, the high thermal activity and the sweeping solar wind which can remove as well as contribute particles.

The Lunar Ionosphere Detector will help identify the ionized charged elements and molecules. It will also measure the charged particles as the Moon passes through the Earth's magnetic field. The experiment is also designed to give us a preliminary value for the electric field of the lunar surface.

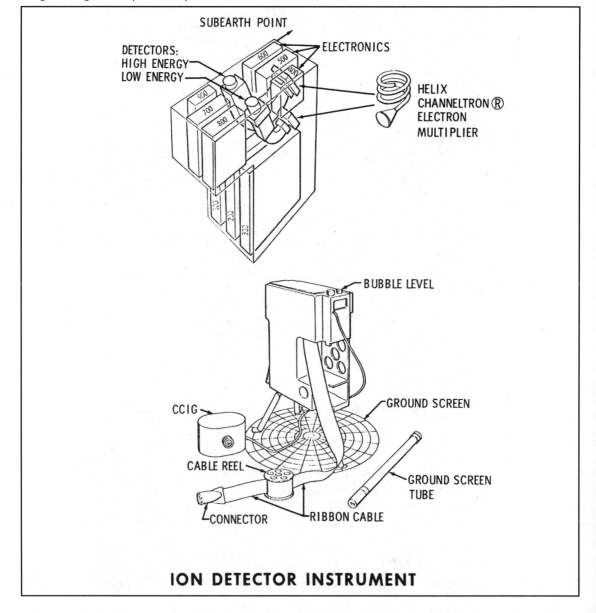

ION DETECTOR INSTRUMENT

The ALSEP Lunar Ionosphere Detector design and subsequent data analysis are the responsibilities of Dr. John Freeman and Dr. Curt Michel both of Rice University. The LID utilizes a suprathermal ion detector to detect and count positive ions. It contains two curved plate analyzers that measure the energy of positive ions. One curved plate analyzer measures ions with an energy range from 0.2 electron volts to 48.6 (e.v.). It contains a velocity filter (crossed magnetic and electric fields) that admits ions with velocities from 4×10^4 to 9.35×10^6 cm/sec. From the velocity data, the mass of the solar wind particles in the energy range from 10 e.v. to 3500 e.v. can be determined. The mass of these high energy particles from solar flares cannot be determined because the analyzer does not have the velocity selector.

The weight of the instrument is 19.6 pounds, operational power is 60 watts, and input voltage is +29 VDC.

Lunar Atmosphere Detector (LAD) — The scientific objective of the Lunar Atmosphere Detector is to measure the density, temperature, and the changes in the lunar atmosphere. The LAD basically measures the total pressure of neutral (inactive) particles whereas the Ionosphere Detector measures composition of the ionized (active) particles. The densities are expected to build up on the sunlit side and to fall off on the dark side of the Moon. These measurements are expected to contribute to understanding of the processes which shape the lunar surface. The erosional features, recognized from the Apollo 11 soil samples, are produced by forces which are not similar to processes known on Earth.

The design of the experiment and the subsequent data analysis are the responsibility of Dr. Francis Johnson, Southwest Center for Advanced Studies and Mr. Dallas Evans, NASA Manned Spacecraft Center.

The lunar atmosphere detector consists of a cold cathode ion gauge assembly, electronics package, and structural and thermal housings. The neutral particles are ionized and collected by the cathode, which is one of a pair of sensor electrodes, and produce a current at the input circuitry of the electronics proportional to the particle density. This signal is amplified and processed by the electronics for transmission through the central station back to Earth. The gauge temperature is measured directly and is inserted into the data handling circuitry.

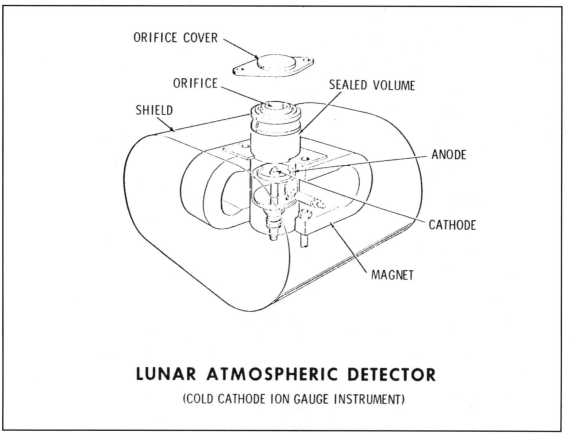

LUNAR ATMOSPHERIC DETECTOR
(COLD CATHODE ION GAUGE INSTRUMENT)

Any one of seven different dynamic ranges may be selected, permitting detection of neutral atom densities ranging from 10^{-6} torr to 10^{-12} torr. The desired sensitivity may be chosen by command from Earth or by internal adjustments. This experiment will give the particle density and temperature of the ambient lunar atmosphere, including any temporal variations either of a random character or associated with lunar local time or solar activity.

Instrument parameters are as follows:

Gauge accuracy	+ 30% above 10^{-10} torr
Weight	+ 50% below 10^{-10} torr
Operating Power	2 watts

The Solar Wind Spectrometer — The Solar Wind Spectrometer will measure the strength, velocity and directions of the electrons and protons which emanate from the Sun and reach the lunar surface. The solar wind is the major external force working on the Moon's surface. The spectrometer measurements will help interpret the magnetic field of the Moon, the lunar atmosphere and the analysis of lunar samples.

Knowledge of the solar wind will help us understand the origin of the Sun and the physical processes at work on the Sun, i.e. the creation and acceleration of these particles and how they propagate through interplanetary space. It has been calculated that the solar wind puts one kiloton of energy into the Earth's magnetic field every second. This enormous amount of energy influences such Earth's processes as the aurora, ionosphere and weather. Although it requires twenty minutes for a kiloton to strike the Moon its effects should be apparent in many ways.

In addition to the Solar Wind Spectrometer, an independent experiment (the Solar Wind Composition Experiment) will collect the gases of the solar wind for return to Earth for analysis.

The design of the spectrometer and the subsequent data analysis are the responsibility of Dr. Conway Snyder, Dr. Douglas Clay and Dr. Marcia Neugebauer of the Jet Propulsion Laboratory.

Seven identical modified Faraday cups (an instrument that traps ionized particles) are used to detect and collect solar wind electrons and protons. One cup is to the vertical, whereas the remaining six cups surround the vertical where the angle between the normals of any two adjacent cups is approximately 60 degrees. Each cup measures the current produced by the charged particle flux entering into it. Since the cups are identical, and if the particle flux is equal in each direction, equal current will be produced in each cup. If the flux is not equal in each direction, analysis of the amount of current in the seven cups will determine the variation of particle flow with direction. Also, by successively changing the voltages on the grid of the cup and measuring the corresponding current, complete energy spectra of both electrons and protons in the solar wind are produced.

Data from each cup is processed in the ALSEP data subsystem. The measurement cycle is organized into 16 sequences of 186 ten-bit words. The instrument weighs 12.5 pounds, has an input voltage of about 28.5 volts and has an average input power of about 3.2 watts. The measurement ranges are as follows:

Electrons

	High gain modulation	10.5 - 1376 e.v. (electron volts)
	Low gain modulation	6.2 - 817 e.v.

Protons

	High gain modulation	75 - 9600 e.v.
	Low gain modulation	45 - 5700 e.v.

Field of View	6.0 Steradians
Angular Resolution	15° (approximately)
Minimum Flux Detectable	10^6 particles/cm^2/sec

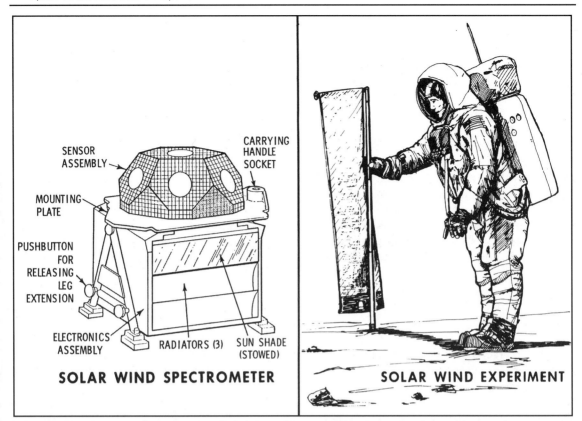

SENSOR ASSEMBLY
CARRYING HANDLE SOCKET
MOUNTING PLATE
PUSHBUTTON FOR RELEASING LEG EXTENSION
ELECTRONICS ASSEMBLY
RADIATORS (3)
SUN SHADE (STOWED)

SOLAR WIND SPECTROMETER

SOLAR WIND EXPERIMENT

Solar Wind Composition Experiment (SWCE) — The scientific objective of the solar wind composition experiment is to determine the elemental and isotopic composition of the noble gasses in the solar wind. (This is not an ALSEP experiment.)

The solar wind composition detector experiment design and subsequent data analysis are the responsibility of: J. Geiss and P. Eberhardt., University of Bern (Switzerland); P. Signer, Swiss Federal Institute of Technology; with Professor Geiss assuming the responsibility of Principal Investigator.

As in Apollo 11, the SWC detector will be deployed on the Moon and brought back to Earth by the astronauts. The detector, however, will be exposed to the solar wind flux for seventeen hours instead of two hours as in Apollo 11. Also, in the Apollo 11 mission the detector was found to be too close to the working areas of the astronauts. In Apollo 12, the detector will be placed a sufficient distance away from the LM so that it will be free of lunar dust kicked up by astronaut activity.

The solar wind composition detector consists of an aluminum foil 4 ft^2 area and about 0.5 mils thick rimmed by Teflon for resistance to tear during deployment. A staff and yard arrangement will be used to deploy the foil and to maintain the foil approximately perpendicular to the solar wind flux. Solar wind particles will penetrate into the foil, allowing cosmic rays to pass right through. The particles will get firmly trapped at a depth of several hundred atomic layers. After exposure on the lunar surface, the foil is reeled and returned to Earth.

Field Geology Investigations

The scientific objectives of the Apollo Field Geology Investigations are to determine the composition of the Moon and the processes which shape its surfaces. This information will help to determine the history of the Moon and its relationship to the Earth. The early investigations to understanding the nature and origin of the Mare are limited by mission constraints. Apollo 11 visited the Sea of Tranquillity (Mare Tranquillitatis), Apollo 12 will study the Ocean of Storms (Oceanus Procellarum). The results of these studies should help establish the nature of Mare-type areas.

Geology investigations of the Moon actually began with the telescope. Systematic geology mapping began ten years ago with a team of scientists at the U.S. Geological Survey. Ranger, Surveyor and especially Lunar Orbiter data enormously increased the detail and accuracy of these studies. The Apollo 11 investigations represent another enormous advancement in providing new evidence on the Moon's great age, its curious chemistry, the surprisingly high density of the lunar surface material.

On Apollo 12, almost the entire second EVA will be devoted to the Field Geology Investigations and the collection of documented samples. The sample locations will be photographed before and after sampling. The astronauts will carefully describe the setting from which the sample is collected. Samples will be taken along the rays of large craters. It is this material, ejected from great depth, which will provide evidence on the nature of the lunar interior. In addition to specific tasks, the astronauts will be free to photograph and sample phenomena which they judge to be unusual, significant and interesting. The astronauts are provided with a package of detailed photo maps which they will use for planning traverses. Photographs will be taken from the LM window. Each feature or family of features will be described, relating to features on the photo maps. Areas and features where photographs should be taken and representative samples collected will be marked on the maps as determined primarily by the astronauts but with inputs from Earth-based geologists.

The Earth-based geologists will be available to advise the astronauts in real-time and will work with the data returned, the photos, the samples of rock and the astronauts' observations to reconstruct here on Earth the astronauts' traverse on the Moon.

If landing accuracy permits, the Apollo 12 astronauts plan to visit the Surveyor III spacecraft. Analytical results of lunar samples collected from the Surveyor III site will be compared to chemical analysis made by the Surveyor alpha particle back-scatter experiment.

The Field Geological Investigations are the responsibility of Dr. Eugene Shoemaker, Principal Investigator., California Institute of Technology. His Co-Investigators are Aaron Waters, University of California (Santa Cruz); E. M. Goddard., University of Michigan; H. H. Schmitt, Astronaut; T. H. Foss, NASA; J. J. Rennilson, Jet Propulsion Laboratory; Gordon Swann. USGS; M. H. Hait, USGS; E. H. Holt, USGS; and R. M. Batson, USGS.

Each astronaut will carry a Lunar Surface Camera (a modified 70 mm electric Hasselblad). The camera has a 60 mm Biogon lens, with apertures ranging from f/5.6 to f/45. Its focus range is from 3 ft to infinity with detents at the 5 foot, 15 foot and 74 foot settings. The camera system incorporates a rigidly installed glass plate bearing a reference grid immediately in front of the image plane. A polarizing filter attached to the lens of one of the cameras can be rotated in 45° increments for light polarizing studies. On the first EVA, each magazine will carry 160 frames of color film. For the second EVA, each film magazine will contain 200 frames of thin-base black and white film.

A gnomon, used for metric control of near field (less than 10 feet) stereoscopic photography, will provide angular orientation relative to the local vertical. Information on the distances to objects and on the pitch, roll, and azimuth of the camera's optic axis are thereby included in each photograph. The gnomon is a weighted tube suspended vertically on a tripod supported gimbal. The tube extends one foot above the gimbal and is painted with a gray scale in bands one centimeter wide. Photogrammetric techniques will be used to produce three-dimensional models and maps of the lunar surface from the angular and distance relationship between specific objects recorded on the film.

The Apollo black and white surface television camera has two resolution modes (320 scan lines/frame and 1280 scan lines/frame) and two respective scanning modes (10 frames/second and 0.625 frames/second). With the TV camera mounted on a tripod on the lunar surface, the astronauts will be able to conduct the early portion of their traverse within the field of view of the lunar daytime lens. This surveillance will permit Earth-bound advisors to assist in any update of pre-mission plans for the lunar surface operations as such assistance is required.

The 16 mm Data Acquisition Camera will provide time-sequence coverage from within the LM. It can be operated in several automatic modes, ranging from 1 frame/second to 24 frames/sec. Shutter speeds, which are independent of the frame rates, range from 1/1000 second to 1/60 second. Time exposures are also possible. While a variety of lenses is provided, the 18 mm lens will be used to record most of the geological activities in the 1 frame/sec mode.

The Lunar Surface Close-up Camera will be used to obtain high resolution stereoscopic photographs of the lunar surface to provide fine scale information on lunar soil and rock textures. Up to 100 stereo pairs can be exposed on the preloaded roll of 35 mm color film. The handle grip enables the astronaut to operate the camera from a standing position. The film drive and electronic flash are battery operated. The camera photographs a 3"x3" area of the lunar surface.

Geological sampling equipment includes tongs, scoop, hammer, and core tubes. A 24-inch extension handle is provided for several of the tools to aid the astronaut in using them without kneeling.

Sample return containers (SRC) have been provided for return of up to 40 pounds each of lunar material for Earth-based analysis. The SRC's are identical to the ones used on the Apollo 11 mission. They are machined from aluminum forgings and are designed to maintain an internal vacuum during the outbound and return flights. The SRC's will be filled with representative samples of lunar surface material, collected and separately bagged by the astronauts on their traverse and documented by verbal descriptions and photography. Subsurface samples will be obtained by using drive tubes 16 inches long and one inch in diameter. A few grams of material will be preserved under lunar vacuum conditions in a special environmental sample container. This container will be opened for analysis under vacuum conditions equivalent to that at the lunar surface.

Passive Seismic Experiment (PSE) — The ALSEP Passive Seismic Experiment (PSE) will measure seismic activity of the Moon and obtain information on the physical properties of the lunar crust and interior. The PSE will detect surface tilt produced by tidal deformations, moonquakes, and meteorite impacts.

The passive seismometer design and subsequent experiment analysis are the responsibility of Dr. Gary Latham and Dr. Maurice Ewing — Lamont Doherty Geological Observatory; Dr. George Sutton — University of Hawaii; and Dr. Frank Press MIT.

A similar passive seismic experiment, deployed during the Apollo 11 flight, utilized solar energy to produce the power necessary for its operation. Thus, it operated only during the lunar day. The instrument for Apollo 12 utilizes nuclear power and can operate continuously. On Apollo 11, an electronics package served as the base for the seismometer, somewhat isolating the instrument from the lunar surface. The ALSEP instrument for Apollo 12 sits on a leveling stool which provides better contact with the lunar surface. The Apollo 11 seismometer had its own self-contained electronics and transmitter. The Apollo 12 instrument sends its sensor readings to the ALSEP central station which combines the inputs from all the ALSEP experiments into the proper format and transmits the data back to Earth. False signals should be reduced by physically separating the seismometer from the electronics.

After the two astronauts rejoin the command module, the LM ascent stage will be jettisoned toward the lunar surface impacting approximately 5 nautical miles south of the previously emplaced ALSEP Passive Seismometer. This will provide a calibrated seismic event equivalent to one ton of TNT.

There are three major physical components of the PSE:

The sensor assembly consists of three long-period seismometers with orthogonally oriented, capacitance type seismic sensors, measuring along two horizontal axes and one vertical axis, This is mounted on a gimbal platform assembly. There is one short period seismometer which has magnet type sensors. It is located directly on the base of the sensor assembly.

The leveling stool allows manual leveling of the sensor assembly by the astronaut to within ±5°, and final leveling to within 3 arc seconds by control motors. The thermal shroud covers and helps stabilize the temperature of the sensor assembly. Also, two radioisotope heaters will protect the instrument from the extreme cold of the lunar night.

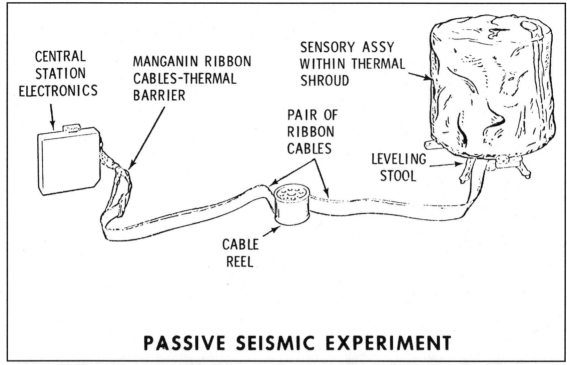

PASSIVE SEISMIC EXPERIMENT

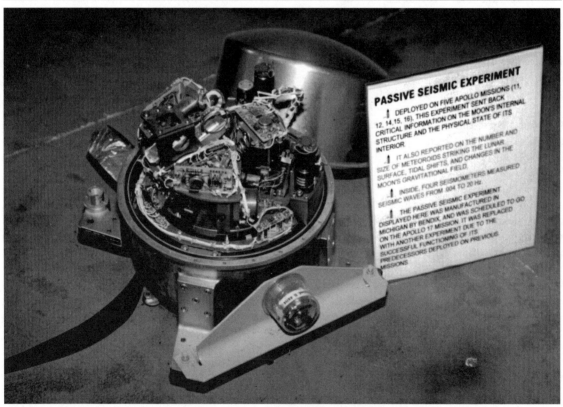

Passive Seismic Experiment on display at the Michigan Space & Science Center.

Dust Detector — The ALSEP Dust Detector is an engineering measurement designed to detect the presence of dust or debris that may impinge on the ALSEP or accumulate during its operating life.

The measurement apparatus consists of three calibrated solar cells, one pointing in east, west and vertical to face the elliptic path of the Sun. The detector is located on the central station.

Dust accumulation on the surface of the three solar cells will reduce the solar illumination detected by the cells. The temperature of each cell will be measured and compared with predicted values.

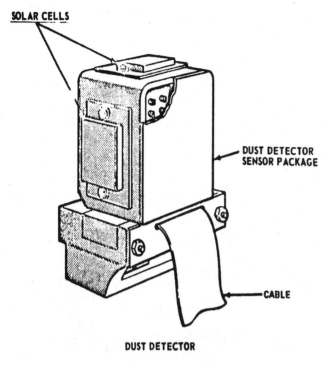

SOLAR CELLS

DUST DETECTOR SENSOR PACKAGE

CABLE

DUST DETECTOR

Lunar Multispectral Photography Experiment S-158 — The objective of the Lunar Multispectral Photography Experiment is to photograph the lunar surface from orbit at four widely separated wavelengths in the green, blue, red and infrared portions of the spectrum. Four 80mm Hasselblad cameras each with a different filter are to be mounted in a ring attached to the command module hatch window. Black and white film will be used in each camera.

Photography will be carried out during the 27th and 28th lunar orbits by astronaut Richard Gordon while alone in the command module. An automatic device will trip the camera shutters simultaneously at 20 second intervals. Vertical strip photography and photographs of possible future landing sites are planned.

The returned film will be analyzed by both photographic and computer methods to produce specially enhanced color composite prints designed to reveal, at high resolution, subtle color shading on the lunar surface that cannot be perceived by the eye or seen on normal color film. There is good evidence that these color differences are related to compositional variations. The enhanced pictures will aid geologists in planning for future sample collection and aid in extrapolating known compositions from returned samples to other parts of the Moon which will not be visited by man.

BOOTSTRAP PHOTOGRAPHY

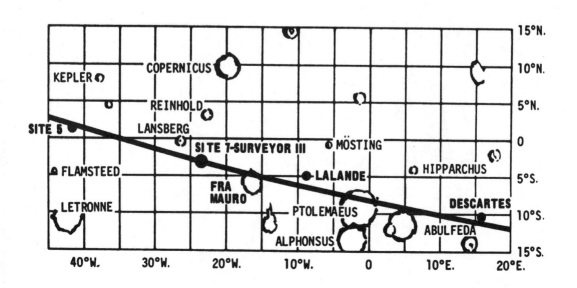

The principal investigator is Dr. Alexander F. H. Goetz/ Bellcomm, Inc.. Co-investigators are Mr. Fred C. Billingsley/ Jet Propulsion Laboratory Dr. Thomas B. McCord/Massachusetts Institute of Technology and Dr. Edward Yost/Long Island University.

SNAP-27 — SNAP-27 is one of a series of radioisotope thermoelectric generators, or atomic batteries, developed by the Atomic Energy Commission under its SNAP program. The SNAP (Systems for Nuclear Auxiliary Power) program is directed at development of generators and reactors for use in space, on land and in the sea.

While nuclear heaters were used in the seismometer package on Apollo 11, SNAP-27 on Apollo 12 will mark the first use of a nuclear electrical power system on the Moon. It is designed to provide all the electricity for continuous one-year operation of the NASA Apollo Lunar Surface Experiments Package.

A nuclear power generator of different design is providing part of the power, along with solar cells, for the Nimbus III satellite which was launched in April 1969. Nimbus III represents the first use of a nuclear power system on a NASA spacecraft. Other systems of this type have also been used on Department of Defense navigational satellites. Altogether eight nuclear power systems, before SNAP-27, have been launched in the United States space program.

The basic SNAP-27 unit is designed to produce at least 63 electrical watts of power for the Apollo 12 lunar surface experiments. It is a cylindrical generator, fueled with the radioisotope Plutonium 238. It is about 18 inches high and 16 inches in diameter, including the heat radiating fins. The generator, making maximum use of the lightweight material beryllium, weighs about 28 pounds unfueled.

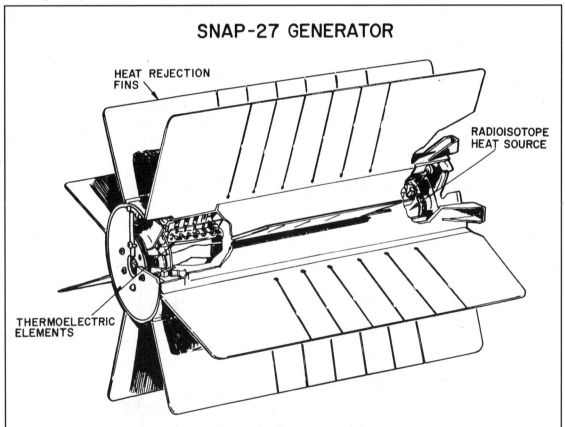

SNAP-27 GENERATOR

HEAT REJECTION FINS

RADIOISOTOPE HEAT SOURCE

THERMOELECTRIC ELEMENTS

The fuel capsule, made of a super-alloy material, is 16.5 inches long and 2.5 inches in diameter. It weighs about 15.5 pounds, of which 8.36 pounds represent fuel.

The Plutonium 238 fuel is fully oxidized and is chemically and biologically inert.

The rugged fuel capsule is contained within a graphite fuel cask from launch through lunar landing. The cask is designed to provide reentry heating protection and added containment for the fuel capsule in the unlikely event of an aborted mission. The cylindrical cask with hemispherical ends includes a primary graphite heat shield, a secondary beryllium thermal shield, and a fuel capsule support structure made of titanium and Inconel materials. The cask is 23 inches long and 8 inches in diameter and weighs about 24.5 pounds. With the fuel capsule installed, it weighs about 40 pounds. It is mounted on the lunar module descent stage by a titanium support structure.

Once the lunar module is on the Moon, an Apollo astronaut will remove the fuel capsule from the cask and insert it into the SNAP-27 generator which will have been placed on the lunar surface near the module.

The spontaneous radioactive decay of the Plutonium 238 within the fuel capsule generates heat in the generator. An assembly of 442 lead telluride thermoelectric elements converts this heat 1480 thermal watts — directly into electrical energy — at least 63 watts. There are no moving parts.

The unique properties of plutonium 238 make it an excellent isotope for use in space nuclear generators. At the end of almost 90 years, plutonium 238 is still supplying half of its original heat. In the decay process, plutonium 238 emits mainly the nuclei of helium (alpha radiation), a very mild type of radiation with a short emission range.

Before the use of the SNAP-27 system in the Apollo program was authorized, a thorough review was conducted to assure the health and safety of personnel involved in the launch and of the general public. Extensive safety analyses and tests were conducted which demonstrated that the fuel would be safely contained under almost all credible accident conditions.

Contractors for SNAP-27

General Electric Co., Missile and Space Division., Philadelphia., Pa., designed, developed and fabricated the SNAP-27 generator for the ALSEP.

The 3M Co., St. Paul., Minn, fabricated the thermoelectric elements and assembled the SNAP-27 generator.

Solar Division of International Harvester, San Diego, Calif., fabricated the generator's beryllium structure.

Hitco, Gardena, Calif., fabricated the graphite structure for the SNAP-27 Graphite LM Fuel Cask.

Sandia Corporation, a subsidiary of Western Electric operator of AEC's Sandia Laboratory, Albuquerque., N.M., provided technical direction for the SNAP-27 program.

Savannah River Laboratory, Aiken, South Carolina, operated by the DuPont Company for the AEC, prepared the raw plutonium fuel.

Mound Laboratory, Miamisburg, Ohio, operated by Monsanto Research Corp., for the AEC, fabricated the raw fuel into the final fuel form and encapsulated the fuel.

PHOTOGRAPHIC EQUIPMENT

Still and motion pictures will be made of most spacecraft maneuvers, crew lunar surface activities, and mapping photos from orbital altitude to aid in planning future landing missions. During lunar surface activities, emphasis will be on photographic documentation of lunar surface features and lunar material sample collection.

Camera equipment stowed in the Apollo 12 command module consists of one 70mm Hasselblad electric camera for general photography, the four-camera lunar multispectral camera assembly for the S-158 experiment, and a 16mm motion picture camera. The S-158 experiment camera group consists of four

Hasselblads side-by-side on a common mount, each fitted with a different filter and type of film. A similar experiment was flown as S-065 Earth multispectral photography on Apollo 9.

S-158 experiment objectives are: gathering of lunar surface color variation data for geologic mapping, correlation of photos with spectral reflectance of returned samples as a possible means of determining surface composition with orbital photography, potential landing site photography (Fra Mauro, Descartes, LaLande), and comparison of lunar reflectance variation and wavelengths.

Cameras stowed on the lunar module are two 70mm Hasselblad data cameras fitted with 60mm Zeiss Metric lens, a 16mm motion picture camera fitted with a 10mm lens, and a close-up stereo camera for geological photos on the lunar surface which is stowed in the MESA on the LM descent stage. The LM Hasselblads have crew chest mounts that leave both hands free.

The command module Hasselblad electric camera is normally fitted with an 80mm f/2.8 Zeiss Planar lens, but bayonet mount 250mm and 500mm lens may be substituted for special tasks. The 80mm lens has a focussing range from three feet to infinity and has a field of view of 38 degrees vertical and horizontal on the square-format film frame. Accessories for the command module Hasselblad include a spotmeter, intervalometer, remote control cable, and film magazines. Hasselblad shutter speeds range from time exposure and one second to 1/500 second.

The Maurer 16mm motion picture camera in the command module has lenses of 5, 18 and 75mm available. The camera weighs 2.8 pounds with a 130-foot film magazine attached. Accessories include a right-angle mirror, a power cable, and a sextant adapter which allows the camera to use the navigation sextant optical system. This camera will be mounted in the right-hand window to record descent and landing and the two EVA periods.

The 35mm stereo close-up camera stowed in the LM MESA shoots 24mm square stereo pairs with an image scale of one-half actual size. The camera is fixed focus and is equipped with a standoff hood to position the camera at the proper focus distance. A long handle permits an EVA crewman to position the camera without stooping for surface object photography. Detail as small as 40 microns can be recorded. The camera allows photography of significant surface structure which would remain intact only in the lunar environment, such as fine powdery deposits, cracks or holes, and adhesion of particles. A battery-powered electronic flash provides illumination, and film capacity is a minimum of 100 stereo pairs.

LUNAR DESCRIPTION

Terrain — Mountainous and crater-pitted, the former rising as high as 29 thousand feet and the latter ranging from a few inches to 180 miles in diameter. The craters are thought to be formed primarily by the impact of meteorites. The surface is covered with a layer of fine grained material resembling silt or sand, as well as small rocks and boulders.

Environment — No air, no wind, and no moisture. The temperature ranges from 243 degrees F. in the two-week lunar day to 279 degrees below zero in the two-week lunar night. Gravity is one sixth that of Earth. Micrometeoroids pelt the Moon since there is no atmosphere to burn them up. Radiation might present a problem during periods of unusual solar activity.

Far Side — The far or hidden side of the Moon no longer is a complete mystery. It was first photographed by a Russian craft and since then has been photographed many times, particularly from NASA's Lunar Orbiter and Apollo spacecraft.

Origin — There is still no agreement among scientists on the origin of the Moon. The three theories, (1) the Moon once was part of Earth and split off into its own orbit, (2) it evolved as a separate body at the same time as Earth, and (3) it formed elsewhere in space and wandered until it was captured by Earth's gravitational field.

Physical Facts

Diameter	2,160 miles (about ¼ that of Earth)
Circumference	6,790 miles (about ¼ that of Earth)
Distance from Earth	238,857 miles (mean; 221,463 minimum to 252,710 maximum)
Surface temperature	+243° F (Sun at zenith) −279° F (night)
Surface gravity	1/6 that of Earth
Mass	1/100th that of Earth
Volume	1/50th that of Earth
Lunar day and night	14 Earth days each
Mean velocity in orbit	2,287 miles-per-hour
Escape velocity	1.48 miles-per-second
Month (period of rotation around Earth)	27 days, 7 hours, 43 minutes

Apollo 12 Landing Site

The primary landing site for the Apollo 12 is designated Site 7, located in the Ocean of Storms at 2.94° south latitude by 23.45° west longitude — about 830 nm west of Apollo 11's landing site last July in the Sea of Tranquillity.

Should the Apollo 12 launch be delayed beyond November 14 a secondary site at 2° north latitude by 42° west longitude would be targeted for a November 16 launch. The secondary site is designated Site 5.

Apollo 11 landed at Site 2 on July 20, 1969. Actual landing took place at 0°41'15" north latitude by 23°26' east longitude, some 6,870 meters west of the Site 2 ellipse center.

A possible added bonus to a pinpoint Apollo 12 landing will be the Surveyor III spacecraft located on the inner slope of a crater at 3.33° south latitude by 23.17° west longitude 1,118 feet from the Apollo 12 aiming point. Retrieval of Surveyor components exposed to almost three years in the lunar environment is a low-priority objective of Apollo 12, coming after sample collection, EVA operations, ALSEP deployment, and expansion of lunar exploration begun by Apollo 11.

SATURN V LAUNCH VEHICLE

The launch vehicle for the Apollo 12 mission is essentially the same as that for Apollo 11. The number of instrumentation measurements, 1,365 on AS-507 is only 17 more than were taken on the vehicle that launched the spacecraft on the first lunar landing mission.

First Stage

The Marshall Space Flight Center and The Boeing Co. jointly developed the 7.6-million pound thrust first stage (S-IC) of the Saturn V. Major structural components include the forward skirt, oxidizer tank, intertank structure, fuel tank and thrust structure. The normal propellant flow rate to the five F-1 engines is 29,364.5 pounds (2,230 gallons) per second. Four of the engines are mounted on a ring at 90-degree intervals. These four are gimballed to control the rocket's attitude in flight. The fifth engine is mounted rigidly in the center.

Second Stage

The second stage (S-II) is built by the Space Division of North American Rockwell Corp. at Seal Beach, Calif. Its major structural components include the forward skirt, the liquid hydrogen and liquid oxygen tanks (separated by an insulated common bulkhead), the thrust structure and an interstage section that connects the first and second stages.

Four of the stage's five J-2 engines are mounted on a 17.5 foot diameter ring. These four may be gimballed through a plus or minus seven degree pattern for thrust vector control. The fifth engine is mounted rigidly on the stage centerline.

Third Stage

The McDonnell Douglas Astronautics Co. at Huntington Beach, Calif. produces the third stage (S-IVB). Its major structural components include aft interstage and skirt, thrust structure, propellant tanks with common bulkhead, forward skirt, and a single J-2 engine.

Insulation between the stage's propellant tanks is necessary because the liquid oxygen, at about 293 degrees below zero Fahrenheit is warm enough, relatively, to heat the liquid hydrogen, at 423 degrees below zero, and cause excessive vaporization.

The gimbaled J-2 engine that powers the stage is capable of a maximum of 230,000 pounds of thrust. On the Apollo 12 mission the thrust range will be from 176,982 to 207,256 pounds. The S-IVB, capable of shutdown and restart, will provide propulsion twice during the Apollo 12 mission.

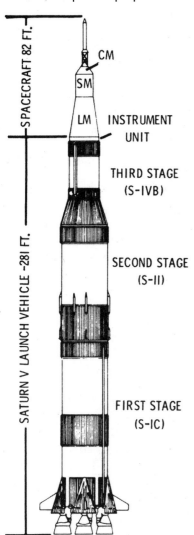

FIRST STAGE (S-IC)

Diameter	33 feet
Height	138 feet
Weight	5,030,715 lbs. fueled 287,850 lbs. dry
Engines	Five F-1
Propellants	Liquid oxygen (3,308,605 lbs., 340,531 gals.)
	RP-1 (kerosene) (1,428,855 lbs., 209,792 gals.)
Thrust	7,620,427 lbs. at liftoff

SECOND STAGE (S-II)

Diameter	33 feet
Height	81.5 feet
Weight	1,060,420 lbs. fueled 80,220 lbs. dry
Engines	Five J-2
Propellants	Liquid oxygen (820,710 lbs., 86,208 gals.)
	Liquid hydrogen (158,230 lbs., 282,543 gals.)
Thrust	1,101,135 to 1,161,414 lbs.
Interstages	11,465

THIRD STAGE (S-IVB)

Diameter	21.7 feet
Height	58.3 feet
Weight	262,070 lbs. fueled, 25,050 lbs. dry
Engine	One J-2
Propellants	Liquid oxygen (289,900 lbs, 19,766 gals.)
	Liquid hydrogen (43,500 lbs., 77,671 gals.)
Thrust	176,982 to 207,256 lbs.
Interstage	8,035 lbs,

INSTRUMENT UNIT

Diameter	21.7 feet
Height	3 feet
Weight	4,275 lbs.

NOTE: Weights and measures given above are for the nominal vehicle configuration for Apollo 12. The figures may vary slightly due to changes before launch to meet changing conditions. Weights of dry stages and propellants do not equal total weight because frost and miscellaneous smaller items are not included in chart.

Instrument Unit

The instrument unit (IU) contains the navigation, guidance and control equipment to steer the vehicle through its Earth orbits and into the final translunar trajectory maneuver. The six major systems are structural, thermal control, guidance and control, measuring and telemetry, radio frequency, and electrical.

In addition to navigation, guidance, and control of the vehicle the instrument unit provides measurement of the vehicle performance and environment; data transmission with ground stations; radio tracking of the vehicle; checkout and monitoring of vehicle functions; initiation of stage functional sequencing; detection of emergency situations; power storage and network distribution of its electric power system; and checkout of preflight, launch and flight functions.

A path-adaptive guidance scheme is used in the Saturn V instrument unit. A programmed trajectory is used during first stage boost with guidance beginning only after the vehicle has left the atmosphere. This prevents movements that might cause the vehicle to break apart while attempting to compensate for winds, jet streams, and gusts encountered in the atmosphere.

If after second stage ignition the vehicle deviates from the optimum trajectory in climb, the vehicle derives and corrects to a new trajectory.

The ST-124M inertial platform — the heart of the navigation, guidance and control system — provides space-fixed reference coordinates and measures acceleration along the three mutually perpendicular axes of the coordinate system. If the inertial platform fails during boost, spacecraft systems continue guidance and control functions for the rocket. After second stage ignition the crew can manually steer the space vehicle.

International Business Machines Corp, is prime contractor for the instrument unit.

Propulsion

The 37 rocket engines of the Saturn V have thrust ratings ranging from 70 pounds to more than 1.5 million pounds, Some engines burn liquid propellants, others use solids.

Engines in the first stage develop approximately 1,524,085 pounds of thrust each at liftoff, building up to about 1,808,508 pounds before cutoff. The cluster of five engines gives the first stage a thrust range of from 7,620,427 pounds at liftoff to 9,042,041 pounds just before center engine cutoff.

The F-1 engine weighs almost 10 tons, is more than 18 feet high and has a nozzle-exit diameter of nearly 14 feet. The engine consumes almost three tons of propellant per second.

The first stage has eight solid-fuel retrorockets which separate the stage from the second stage. Each rocket produces a thrust of 87,900 pounds for 0.6 second.

The second stage engine thrust varies from 220,227 to 232,283 during this flight. The 3,500-pound J-2 is more efficient than the F-1 because it burns the high-energy fuel hydrogen. F-1 and J-2 engines are produced by the Rocketdyne Division of North American Rockwell Corp.

The second stage has four 219,000 pound-thrust solid fuel ullage rockets to settle liquid propellant in the bottom of the main tanks and help attain a "clean" separation from the first stage. Four retrorockets are located in the S-IVB aft interstage (which never separates from the S-II) to separate the S-II from the S-IVB. There are two jettisonable ullage rockets for propellant settling prior to engine ignition. Eight smaller engines in the two auxiliary propulsion system modules on the S-IVB stage provide 3-axis attitude control.

COMMAND AND SERVICE MODULE STRUCTURE, SYSTEMS

The Apollo spacecraft for the Apollo 12 mission is comprised of Command Module 108, Service Module 108, Lunar Module 6, a spacecraft-lunar module adapter (SLA) and a launch escape system. The SLA houses the lunar module and serves as a mating structure between the Saturn V instrument unit and the LM.

Launch Escape System (LES) — Propels the command module to safety in an aborted launch. It has three solid-propellant rocket motors: a 147,000 pound-thrust launch escape system motor, a 2,400-pound-thrust pitch control motor, and a 31,500 pound-thrust tower jettison motor. Two canard vanes deploy to turn the command module aerodynamically to an attitude with the heat-shield forward. The system is 33 feet tall, four feet in diameter at the base, and weighs 8,945 pounds.

Command Module (CM) Structure — The command module is a pressure vessel encased in heat shields, cone-shaped, weighing 12,365 pounds.

The command module consists of a forward compartment which contains two reaction control engines and components of the Earth landing system, the crew compartment or inner pressure vessel containing crew accommodations, controls and displays, and many of the spacecraft systems, and the aft compartment housing ten reaction control engines, propellant tankage, helium tanks, water tanks, and the CSM umbilical cable. The crew compartment contains 210 cubic feet of habitable volume.

Heat-shields around the three compartments are made of brazed stainless steel honeycomb with an outer layer of phenolic epoxy resin as an ablative material.

CSM 108 and LM-6 are equipped with the probe-and-drogue docking hardware. The probe assembly is a powered folding coupling and impact attenuating device mounted in the CM tunnel that mates with a conical drogue mounted in the LM docking tunnel. After the 12 automatic docking latches are checked following a docking maneuver, both the probe and drogue are removed to allow crew transfer between the CSM and LM.

Service Module (SM) Structure — The service module for the Apollo 12 mission will weigh 51,105 pounds. Aluminum honeycomb panels one inch thick form the outer skin, and milled aluminum radial beams separate the interior into six sections around a central cylinder containing two helium spheres, four sections containing service propulsion system fuel-oxidizer tankage, another containing fuel cells, cryogenic oxygen and hydrogen, and one sector essentially empty.

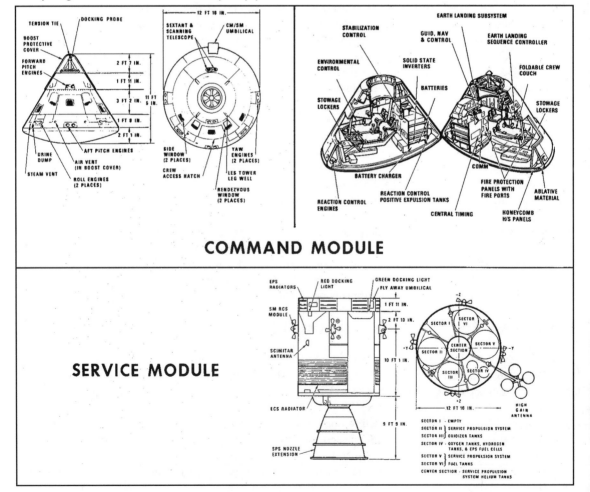

COMMAND MODULE

SERVICE MODULE

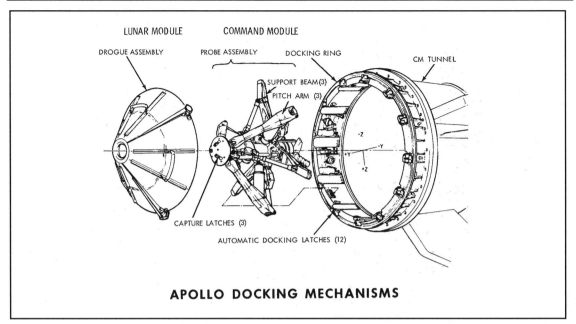

APOLLO DOCKING MECHANISMS

Spacecraft LM Adapter (SLA) Structure — The spacecraft LM adapter is a truncated cone 28 feet long tapering from 260 inches diameter at the base to 154 inches at the forward end at the service module mating line. The SLA weighs 4,000 pounds and houses the LM during launch and Earth orbital flight.

CSM SYSTEMS

Guidance, Navigation and Control System (GNCS) — Measures and controls spacecraft position, attitude, and velocity, calculates trajectory, controls spacecraft propulsion system thrust vector, and displays abort data. The guidance system consists of three subsystems: Inertial, made up of an inertial measurement unit and associated power and data components; computer which processes information to or from other components; and optics consisting of scanning telescope and sextant for celestial and/or landmark sighting for spacecraft navigation. VHF ranging device serves as a backup to the LM rendezvous radar.

Stabilization and Control Systems (SCS) — Controls spacecraft rotation, translation, and thrust vector and provides displays for crew-initiated maneuvers; backs up the guidance system for control functions. It has three subsystems; attitude reference, attitude control, and thrust vector control.

Service Propulsion System (SPS) — Provides thrust for large spacecraft velocity changes through a gimbal-mounted 20,500-pound thrust hypergolic engine using a nitrogen tetroxide oxidizer and a 50-50 mixture of unsymmetrical dimethyl hydrazine and hydrazine fuel. This system is in the service module. The system responds to automatic firing commands from the guidance and navigation system or to manual commands from the crew. The engine thrust level is not throttleable. The stabilization and control system gimbals the engine to direct the thrust vector through the spacecraft center of gravity.

Telecommunications System — Provides voice, television, telemetry, and command data and tracking and ranging between the spacecraft and Earth, between the command module and the lunar module and between the spacecraft and the extravehicular astronaut. It also provides intercommunications between astronauts.

The high-gain steerable S-Band antenna consists of four, 31 inch-diameter parabolic dishes mounted on a folding boom at the aft end of the service module. Signals from the ground stations can be tracked either automatically or manually with the antenna's gimballing system. Normal S-Band voice and uplink/down-link communications will be handled by the omni and high-gain antennas.

Sequential System — Interfaces with other spacecraft systems and subsystems to initiate time critical functions during launch, docking maneuvers, sub-orbital aborts, and entry portions of a mission. The system also controls routine spacecraft sequencing such as service module separation and deployment of the Earth landing system.

Emergency Detection System (EDS) — Detects and displays to the crew launch vehicle emergency conditions, such as excessive pitch or roll rates or two engines out, and automatically or manually shuts down the booster and activates the launch escape system; functions until the spacecraft is in orbit.

Earth Landing System (ELS) — Includes the drogue and main parachute system as well as postlanding recovery aids. In a normal entry descent, the command module forward heat shield is jettisoned at 24,000 feet, permitting mortar deployment of two reefed 16.5-foot diameter drogue parachutes for orienting and decelerating the spacecraft. After disreef and drogue release, three mortar deployed pilot chutes pull out the three main 83.3 foot diameter parachutes with two-stage reefing to provide gradual inflation in three steps. Two main parachutes out of three can provide a safe landing.

Reaction Control System (RCS) — The SM RCS has four identical RCS "quads" mounted around the SM 90 degrees apart. Each quad has four 100 pound-thrust engines, two fuel and two oxidizer tanks and a helium pressurization sphere. Attitude control and small velocity maneuvers are made with the SM RCS.

The CM RCS consists of two independent six-engine subsystems of six 93 pound-thrust engines each used for spacecraft attitude control during entry. Propellants for both CM and SM RCS are monomethyl hydrazine fuel and nitrogen tetroxide oxidizer with helium pressurization. These propellants burn spontaneously when combined (without an igniter).

Electrical Power System (EPS) — Provides electrical energy sources, power generation control, power conversion and conditioning, and power distribution to the spacecraft. The primary source of electrical power is the fuel cells mounted in the SM. The fuel cell also furnishes drinking water to the astronauts as a by-product of the fuel cells.

Three silver-zinc oxide storage batteries supply power to the CM during entry and after landing, provide power for sequence controllers, and supplement the fuel cells during periods of peak power demand. A battery charger assures a full charge prior to entry.

Two other silver-zinc oxide batteries supply power for explosive devices for CM/SM separation, parachute deployment and separation, third-stage separation, launch escape system tower separation, and other Pyrotechnic uses.

Environmental Control System (ECS) — Controls spacecraft atmosphere, pressure, and temperature and manages water. In addition to regulating cabin and suit gas pressure, temperature and humidity, the system removes carbon dioxide, odors and particles, and ventilates the cabin after landing. It collects and stores fuel cell potable water for crew use, supplies water to the glycol evaporators for cooling, and dumps surplus water overboard through the waste H_2O dump nozzle. Proper operating temperature of electronics and electrical equipment is maintained by this system through the use of the cabin heat exchangers, the space radiators, and the glycol evaporators.

Recovery Aids — Recovery aids include the uprighting system, swimmer Interphone connections, sea dye marker, flashing beacon, VHF recovery beacon, and VHF transceiver. The uprighting system consists of three compressor-inflated bags to upright the spacecraft if it should land in the water apex down (stable II position).

Caution and Warning System — Monitors spacecraft systems for out-of-tolerance conditions and alerts crew by visual and audible alarms.

Controls and Displays — Provide status readouts and control functions of spacecraft systems in the command and service modules. All controls are designed to be operated by crewmen in pressurized suits. Displays are grouped by system and located according to the frequency of use and crew responsibility

<u>Command and Service Module Modifications</u> — Differences between the Apollo 12 CSM 108 and CSM 107 flown on Apollo 11 are as follows:

* Arc suppression networks have been added at each SM RCS engine to protect CSM electronic systems from electromagnetic interference produced by RCS heater cycling.

* Hydrogen separator has been added to water subsystem to prevent hydrogen gas from entering potable water tank. (CSM 107 had the hydrogen separator mounted on the hand water dispenser in the cabin.)

* The S-Band squelch override switch has been moved from the lower equipment bay to the display and control panel for easier crew access.

* The recovery loop for spacecraft retrieval has been strengthened to obviate the requirement for the swimmer to install an auxiliary loop.

LUNAR MODULE STRUCTURES, WEIGHT

The lunar module is a two-stage vehicle designed for space operations near and on the Moon. The lunar module stands 22 feet 11 inches high and is 31 feet wide (diagonally across landing gear). The ascent and descent stages of the LM operate as a unit until staging, when the ascent stage functions as a single spacecraft for rendezvous and docking with the CM.

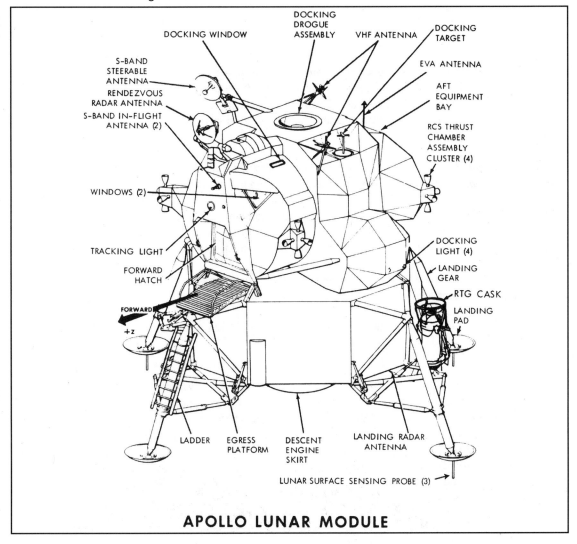

APOLLO LUNAR MODULE

Ascent Stage

Three main sections make up the ascent stage: the crew compartment, midsection, and aft equipment bay. Only the crew compartment and midsection are pressurized (4.8 psig). The cabin volume is 235 cubic feet (6.7 cubic meters). The stage measures 12 feet 4 inches high by 14 feet 1 inch in diameter.
The ascent stage has six substructural areas: crew compartment, midsection, aft equipment bay, thrust chamber assembly cluster supports, antenna supports and thermal and micrometeoroid shield.

The cylindrical crew compartment is 92 inches (2.35 m) in diameter and 42 inches (1.07 m) deep. Two flight stations are equipped with control and display panels, armrests, body restraints, landing aids, two front windows, an overhead docking window, and an alignment optical telescope in the center between the two flight stations. The habitable volume is 160 cubic feet.

A tunnel ring atop the ascent stage meshes with the command module docking latch assemblies. During docking, the CM docking ring and latches are aligned by the LM drogue and the CSM probe.

The docking tunnel extends downward into the midsection 16 inches (40 cm). The tunnel is 32 inches (0.81 cm) in diameter and is used for crew transfer between the CSM and LM. The upper hatch on the inboard end of the docking tunnel opens inward and cannot be opened without equalizing pressure on both hatch surfaces.

A thermal and micrometeoroid shield of multiple layers of mylar and a single thickness of thin aluminum skin encases the entire ascent stage structure.

Descent Stage

The descent stage center compartment houses the descent engine, and descent propellant tanks are housed in the four square bays around the engine. Quadrant II (Seq bay) contains ALSEP, and Radioisotope Thermoelectric Generator (RTG) externally. Quadrant IV contains the MESA. The descent stage measures 10 feet 7 inches high by 14 feet 1 inch in diameter and is encased in the mylar and aluminum alloy thermal and micrometeoroid shield.

The LM egress platform, or "porch", is mounted on the forward outrigger just below the forward hatch. A ladder extends down the forward landing gear strut from the porch for crew lunar surface operations.

The landing gear struts are explosively extended and provide lunar surface landing impact attenuation. The main struts are filled with crushable aluminum honeycomb for absorbing compression loads. Footpads 37 inches (0.95 m) in diameter at the end of each landing gear provide vehicle support on the lunar surface.

Each pad (except forward pad) is fitted with a 68 inch long lunar surface sensing probe which signals the crew to shut down the descent engine upon contact with the lunar surface.

LM-6 flown on the Apollo 12 mission has a launch weight of 33,325 pounds. The weight breakdown is as follows:

Ascent stage, dry	4,760 lbs.	Includes water
		and oxygen; no
Descent stage, dry	4,875 lbs.	crew
RCS propellants (loaded)	595 lbs.	
DPS propellants (loaded)	17,925 lbs.	
APS propellants (loaded)	5,170 lbs.	
	33,325 lbs.	

LUNAR MODULE SYSTEMS

Electrical Power System — The LM DC electrical system consists of six silver zinc primary batteries — four in the descent stage and two in the ascent stage. Twenty-eight - volt DC power is distributed to all LM systems. AC power (117v 400 Hz) is supplied by two inverters.

Environmental Control System — Consists of the atmosphere revitalization section, oxygen supply and cabin pressure control section, water management, heat transport section, and outlets for oxygen and water re-servicing of the portable life support system (PLSS).

Components of the atmosphere revitalization section are the suit circuit assembly which cools and ventilates the pressure garments, reduces carbon dioxide levels, removes odors, noxious gases and excessive moisture; the cabin recirculation assembly which ventilates and controls cabin atmosphere temperatures; and the steam flex duct which vents to space steam from the suit circuit water evaporator.

The oxygen supply and cabin pressure section supplies gaseous oxygen to the atmosphere revitalization section for maintaining suit and cabin pressure. The descent stage oxygen supply provides descent flight phase and lunar stay oxygen needs, and the ascent stage oxygen supply provides oxygen needs for the ascent and rendezvous flight phase.

Water for drinking, cooling, fire fighting, food preparation, and refilling the PLSS cooling water servicing tank is supplied by the water management section. The water is contained in three nitrogen-pressurized bladder-type tanks, one of 367-pound capacity in the descent stage and two of 47.5-pound capacity in the ascent stage.

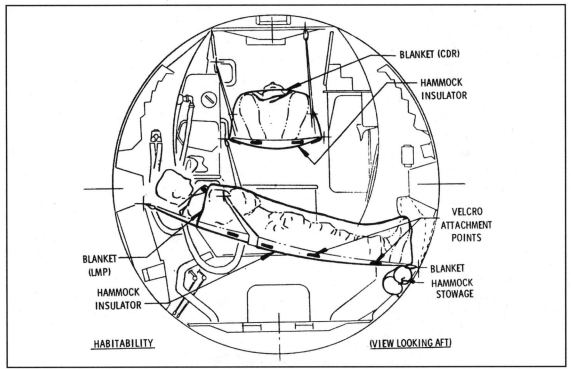

The heat transport section has primary and secondary water-glycol solution coolant loops. The primary coolant loop circulates water-glycol for temperature control of cabin and suit circuit oxygen and for thermal control of batteries and electronic components mounted on cold plates and rails. If the primary loop becomes inoperative, the secondary loop circulates coolant through the rails and cold plates only. Suit circuit cooling during secondary coolant loop operation is provided by the suit loop water boiler. Waste heat from both loops is vented overboard by water evaporation or sublimators.

Crew hammocks and blankets have been provided on LM-6 to give the crew a more comfortable environment for sleep between the EVA periods.

Communications System — Two S-band transmitter-receivers, two VHF transmitter receivers, a signal processing assembly, and associated spacecraft antenna make up the LM communications system. The system transmits and receives voice, tracking and ranging data, and transmits telemetry data on about 270 measurements and TV signals to the ground. Voice communications between the LM and ground stations is by S-band, and between the LM and CSM voice is on VHF.

Although no real-time commands can be sent to the LM, the digital uplink processes guidance officer commands transmitted from Mission Control Center to the LM guidance computer, such as state vector updates. The data storage electronics assembly (DSEA) is a four-channel voice recorder with timing signals with a 10-hour recording capacity which will be brought back into the CSM for return to Earth. DSEA recordings cannot be "dumped" to ground stations.

LM antennas are one 26-inch diameter parabolic S-band steerable antenna, two S-band inflight antennas, two VHF inflight antennas, EVA antenna and an erectable S-band antenna (optional) for lunar surface.

Guidance, Navigation and Control System — Comprised of six sections: primary guidance and navigation, section (PGNS), abort guidance section (AGS), radar section, control electronics section (CES), and orbital rate drive electronics for Apollo and LM (ORDEAL).

* The PGNS is an aided inertial guidance system updated by the alignment optical telescope, an inertial measurement unit, and the rendezvous and landing radars. The system provides inertial reference data for computations, produces inertial alignment reference by feeding optical sighting data into the LM guidance computer, displays position and velocity data, computes LM-CSM rendezvous data from radar inputs, controls attitude and thrust to maintain desired LM trajectory, and controls descent engine throttling and gimbaling.

The LM-6 primary guidance computer has the Luminary 1B Software program, which is an improved version over that in LM-5.

The AGS is an independent backup system for the PGNS, having its own inertial sensors and computer.

* The radar section is made up of the rendezvous radar which provides CSM range and range rate, and line-of-sight angles for maneuver computation to the LM guidance computer; the landing radar which provide altitude and velocity data to the LM guidance computer during lunar landing. The rendezvous radar has an operating range from 80-feet to 400 nautical miles. The ranging tone transfer assembly, utilizing VHF electronics, is a passive responder to the CSM VHF ranging device and is a backup to the rendezvous radar.

* The CES controls LM attitude and translation about all axes. It also controls by PGNS command the automatic operation of the ascent and descent engine and the reaction control thrusters. Manual attitude controller and thrust-translation controller commands are also handled by the CES.

* ORDEAL, displays on the flight director attitude indicator, is the computed local vertical in the pitch axis during circular Earth or lunar orbits.

Reaction Control System — The LM has four RCS engine clusters of four 100-pound (45.4 kg) thrust engines each of which use helium-pressurized hypergolic propellants. The oxidizer is nitrogen tetroxide, fuel is Aerozine 50 (50/50 blend of hydrazine and unsymmetrical dimethyl hydrazine). Interconnect valves permit the RCS system to draw from ascent engine propellant tanks.

The RCS provides small stabilizing impulses during ascent and descent burns, controls LM attitude during maneuvers, and produces thrust for separation, and ascent/descent engine tank ullage. The system may be operated in either the pulse or steady-state modes.

Descent Propulsion System — Maximum rated thrust of the descent engine is 9,870 pounds (4,300.9 kg) and is throttleable between 1,050 pounds (476.7 kg) and 6,300 pounds (2,860.2 kg). The engine can be gimbaled six degrees in any direction in response to attitude commands and compensates for center of gravity offsets. Propellants are helium-pressurized Aerozine 50 and nitrogen tetroxide.

Ascent Propulsion System — The 3,500-pound (1,589 kg) thrust ascent engine is not gimbaled and performs at full thrust. The engine remains dormant until after the ascent stage separates from the descent stage. Propellants are the same as are burned by the RCS engines and the descent engine.

Caution and Warning, Controls and Displays — These two systems have the same function aboard the lunar module as they do aboard the command module (See CSM systems section.)

Tracking and Docking Lights — A flashing tracking light (once per second 20 milliseconds duration) on the front face of the lunar module is an aid for contingency CSM-active rendezvous LM rescue. Visibility ranges from 400 nautical miles through the CSM sextant to 130 miles with the naked eye. Five docking lights analogous to aircraft running lights are mounted on the LM for CSM-active rendezvous: two forward yellow lights, aft white light, port red light and starboard green light. All docking lights have about a 1,000-foot visibility.

Lunar Module Modifications — Differences between the Apollo 12 LM-6 and LM-5 flown on Apollo 11 are as follows:

* The communications system signal processor assembly has been modified to filter out unwanted signals on the intercom circuit and on S-Band backup voice.

* The rendezvous radar antenna has been modified to reduce cyclical range errors.

* In the cabin, the radiation survey meter and bracket have been removed, and the oxygen purge system (OPS) pallet assembly has been modified for direct attachment of the OPS. The descent stage OPS pallet adapter has been eliminated. Provision for two crew sleeping hammocks has been made.

* The descent stage structure has been modified for installation of the Apollo Lunar Surface Experiment Package (ALSEP).

APOLLO 12 CREW

Life Support Equipment - Space Suits

Apollo 12 crewmen will wear two versions of the Apollo space suit: an Intravehicular pressure garment assembly worn by the command module pilot and the extravehicular pressure garment assembly worn by the commander and the lunar module pilot. Both versions are basically identical except that the extravehicular version has an integral thermal/ meteoroid garment over the basic suit.

From the skin out, the basic pressure garment consists of a nomex comfort layer, a neoprene-coated nylon pressure bladder and a nylon restraint layer. The outer layers of the intravehicular suit are, from the inside out, nomex and two layers of Teflon-coated Beta cloth. The extravehicular integral thermal/meteoroid cover consists of a liner of two layers of neoprene-coated nylon, seven layers of Beta/Kapton spacer laminate, and an outer layer of Teflon-coated Beta fabric.

The extravehicular suit, together with a liquid cooling garment, Portable life support system (PLSS), oxygen purge system, lunar extravehicular visor assembly and other components make up the extravehicular mobility-unit (EMU). The EMU provides an extravehicular crewman with life support for a four hour mission outside the lunar module without replenishing expendables. EMU total weight is 183 pounds. The intravehicular suit weighs 35.6 pounds.

Liquid Cooling Garment — A knitted nylon spandex garment with a network of plastic tubing through which cooling water from the PLSS is circulated. It is worn next to the skin and replaces the constant wear garment during EVA only.

Portable Life Support System — A backpack supplying oxygen at 3.7 Psi and cooling water to the liquid cooling garment. Return oxygen is cleansed of solid and gas contaminants by a lithium hydroxide canister. The PLSS includes communications and telemetry equipment displays and controls, and a main power supply. The PLSS is covered by a thermal insulation jacket. (Two stowed in LM).

Oxygen purge system — Mounted atop the PLSS, the oxygen purge system provides a contingency 30-minute supply of gaseous oxygen in two two-pound bottles pressurized to 5,880 psia. The system may also be worn separately on the front of the pressure garment assembly torso. It serves as a mount for the VHF antenna for the PLSS. (Two stowed in LM).

EXTRAVEHICULAR MOBILITY UNIT

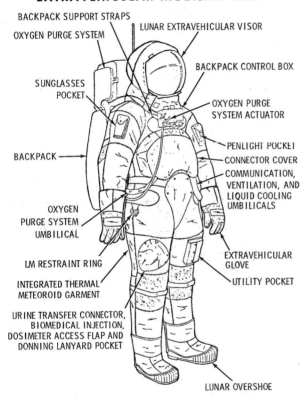

BACKPACK SUPPORT STRAPS
OXYGEN PURGE SYSTEM
LUNAR EXTRAVEHICULAR VISOR
BACKPACK CONTROL BOX
SUNGLASSES POCKET
OXYGEN PURGE SYSTEM ACTUATOR
BACKPACK
PENLIGHT POCKET
CONNECTOR COVER
COMMUNICATION, VENTILATION, AND LIQUID COOLING UMBILICALS
OXYGEN PURGE SYSTEM UMBILICAL
LM RESTRAINT RING
EXTRAVEHICULAR GLOVE
INTEGRATED THERMAL METEOROID GARMENT
UTILITY POCKET
URINE TRANSFER CONNECTOR, BIOMEDICAL INJECTION, DOSIMETER ACCESS FLAP AND DONNING LANYARD POCKET
LUNAR OVERSHOE

Lunar extravehicular visor assembly — A polycarbonate shell and two visors with thermal control and optical coatings on them. The EVA visor is attached over the pressure helmet to provide impact, micrometeoroid, thermal and ultraviolet-infrared light protection to the EVA crewmen.

Extravehicular Gloves — Built of an outer shell of Chromel-R fabric and thermal insulation to provide protection when handling extremely hot and cold objects. The finger tips are made of silicone rubber to provide more sensitivity.

A one-piece constant-wear garment, similar to "long johns" is worn as an undergarment for the space suit in intravehicular operations and for the inflight coveralls. The garment is porous knit cotton with a waist-to-neck zipper for donning. Biomedical harness attach points are provided.

During periods out of the space suits, crewmen will wear two-piece Teflon fabric inflight coveralls for warmth and for pocket stowage of personal items.

Communications carriers ("Snoopy Hats") with redundant microphones and earphones are worn with the pressure helmet; a lightweight headset is worn with the inflight coveralls.

APOLLO LUNAR HAND TOOLS

Special Environmental Container — The special environmental sample is collected in a carefully selected area and sealed in a special container which will retain a high vacuum. The container is opened in the Lunar Receiving Laboratory where it will provide scientists the opportunity to study lunar material in its original environment.

Extension handle — This tool is of aluminum alloy tubing with a malleable stainless steel cap designed to be used as an anvil surface. The handle is designed to be used as an extension for several other tools and to permit their use without requiring the astronaut to kneel or bend down. The handle is approximately 24 inches long and 1 inch in diameter. The handle contains the female half of a quick disconnect fitting designed to resist compression, tension, torsion, or a combination of these loads. Also incorporated are a sliding T handle at the top and an internal mechanism operated by a rotating sleeve which is used with the aseptic collection device.

Four core tubes — These tubes are designed to be driven or augered into loose gravel, sandy material, or into soft rock such as feather rock or pumice. They are about 15 inches in length and one inch in diameter and are made of aluminum tube. Each tube is supplied with a removeable non-serrated cutting edge and a screw-on cap incorporating a metal-to-metal crush seal which replaces the cutting edge. The upper end of each tube is sealed and designed to be used with the extension handle or as an anvil. Incorporated into each tube is a spring device to retain loose materials in the tube.

Scoops (large and small) — This tool is designed for use as a trowel and as a chisel. The scoop is fabricated primarily of aluminum with a hardened-steel cutting edge riveted on and a nine-inch handle. A malleable stainless steel anvil is on the end of the handle. The angle between the scoop pan and the handle allows a compromise for the dual use. The scoop is used either by itself or with the extension handle.

Sampling Hammer — This tool serves three functions, as a sampling hammer a pick or mattock, and as a hammer to drive the core tubes or scoop. The head has a small hammer face on one end, a broad horizontal blade on the other, and large hammering flats on the sides. The handle is fourteen inches long and is made of formed tubular aluminum. The hammer has on its lower end a quick-disconnect to allow attachment to the extension handle for use as a hoe.

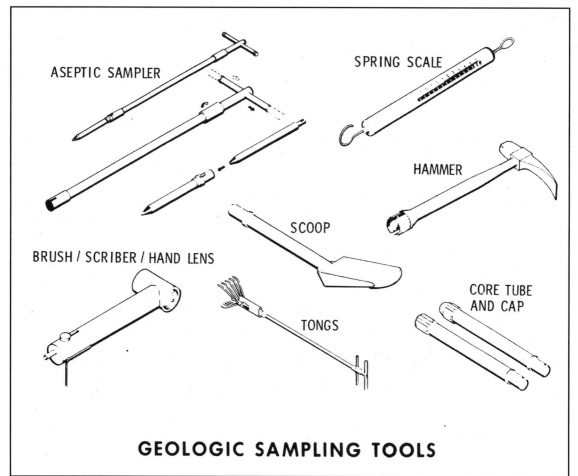

GEOLOGIC SAMPLING TOOLS

Tongs — The tongs are designed to allow the astronaut to retrieve small samples from the lunar surface while in a standing position. The tines are of such angles, length, and number to allow samples of from 3/8 up to 2½-inch diameter to be picked up. This tool is 24 inches in overall length.

Brush/Scriber/Hand Lens — A composite tool

(1)   Brush - To clean samples prior to selection

(2)   Scriber - To scratch samples for selection and to mark for identification

(3)   Hand lens - Magnifying glass to facilitate sample selection

Spring Scale   — To weigh two rock boxes containing lunar material samples, to maintain weight budget for return to Earth.

Instrument staff — The staff holds the Hasselblad camera. The staff breaks down into sections. The upper section telescopes to allow generation of a vertical stereoscopic base of one foot for photography. Positive stops are provided at the extreme of travel. A shaped hand grip aids in aiming and carrying. The bottom section is available in several lengths to suit the staff to astronauts of varying sizes. The device is fabricated from tubular aluminum.

Gnomon — This tool consists of a weighted staff suspended on a two-ring gimbal and supported by a tripod. The staff extends 12 inches above the gimbal and is painted with a gray scale. The gnomon is used as a photographic reference to indicate vertical sun angle and scale. The gnomon has a required accuracy of vertical indication of 20 minutes of arc. Magnetic damping is incorporated to reduce oscillations.

Color Chart — The color chart is painted with three primary colors and a gray scale. It is used as a calibration for lunar photography. The scale is mounted on the tool carrier but may easily be removed and returned to Earth for reference. The color chart is 6 inches in size.

Tool Carrier — The carrier is the stowage container for the tools during the lunar flight. After the landing the carrier serves as support for the astronaut when he kneels down, as a support for the sample bags and samples, and as a tripod base for the instrument staff. The carrier folds flat for stowage. For field use it opens into a triangular configuration. The carrier is constructed of formed sheet metal and approximates a truss structure. Six-inch legs extend from the carrier to elevate the carrying handle sufficiently to be easily grasped by the astronaut.

Field Sample Bags — Approximately 80 bags 4 inches by 5 inches are included in the ALHT for the packaging of samples. These bags are fabricated from Teflon FEP.

Collection Bag — This is a large bag (4 x 8 inches) attached to the astronaut's side of the tool carrier. Field sample bags are stowed in this bag after they have been filled. It can also be used for general storage or to hold items temporarily. (2 in each SRC).

Apollo 12 Crew Menu

More than 70 items comprise the food selection list of freeze-dried rehydratable, wet-pack and spoon-bowl foods. Balanced meals for five days have been packed in man/day overwraps. Items similar to those in the daily menus have been packed in a snack pantry. The snack pantry permits the crew to locate easily a food item in a smorgasbord mode without having to "rob" a regular meal somewhere down deep in a storage box.

Water for drinking and rehydrating food is obtained from two sources in the command module — a dispenser for drinking water and a water spigot at the food preparation station supplying water at about 155 or 55° F. The potable water dispenser squirts water continuously as long as the trigger is held down, and the food preparation spigot dispenses water in one-ounce increments.

A continuous-feed hand water dispenser similar to the one in the command module is used aboard the lunar module for cold-water rehydration of food packets stowed aboard the LM.

After water has been injected into a food bag, it is kneaded for about three minutes. The bag neck is then cut off and the food squeezed into the crewman's mouth. After a meal, germicide pills attached to the outside of the food bags are placed in the bags to prevent fermentation and gas formation. The bags are then rolled and stowed in waste disposal compartments.

The day-by-day, meal-by-meal Apollo 12 Menu for Commander Conrad is on the following page as a typical five-day menu for each crewman.

TYPICAL CREW MENU IS THAT OF APOLLO 12 COMMANDER CONRAD:

APOLLO 12 (CONRAD - RED VELCRO)

NOTE. Supplementary items and meals to this menu are in the spacecraft's Pantry Stowage Section.

MEAL Day 1* 5	Day 2	Day 3	Day 4
A. Peaches IMB	Apricots IMB	Pears IMB	Canadian Bacon
Corn Flakes R	Sausage Patties R	Corn Flakes R	Applesauce RSB
Bacon Squares (8) IMB	Scrambled Eggs RSB	Bacon Squares (8) IMB	Scrambled Eggs RSB
Orange Drink R	Grapefruit Drink R	Grape Drink R	Cinnamon Bread (4) DB
Coffee w/Sugar R	Coffee w/Sugar R	Coffee w/Sugar R	Orange- G.F. Drink R
			Coffee w/Sugar R
B. Tuna Salad RSB	Turkey & Gravy WP	Frankfurters WP	Shrimp Cocktail R
Beef & Gravy WP	Cheese Crackers (4) DB	Applesauce RSB	Ham & Potatoes WP
Jellied Candy IMB	Chocolate Pudding RSB	Chocolate Bar IMB	Apricots IMB
Grape Punch R	Orange-G.F. Drink R	P.A.-G.F. Drink R	Chocolate Pudding RSB
			Orange Drink R
C. Cream of Chicken	Pork & Scalloped	Salmon Salad RSB	Spaghetti w/Meat R
Soup RSB	Potatoes RSB	Chicken Stew RSB	Beef Stew RSB
Chicken & Rice RSB	Bread Slice	Butterscotch	Banana Pudding RSB
Sugar Cookies (4) DB	Sandwich Spread WP	Pudding RSB	Cocoa R
Butterscotch Pudding	Jellied Candy IMB	Peaches IMB	Grape Punch R
RSB	Cocoa R	Grapefruit Drink R	
P.A.-G.F. Drink R	Orange Drink R		
TOTAL CALORIES			
2215	2346	2328	2106

* Day I consists of Meal B and C only.
IMB - Intermediate Moisture Bite, R – Rehydratable, RSB - Rehydratable Spoon-Bowl, WP = Wet Pack, DB = Dry Bite

Personal Hygiene

Crew Personal hygiene equipment aboard Apollo 12 includes body cleanliness items, the waste management system and one-medical kit.

Packaged with the food are a toothbrush and a two-ounce tube of toothpaste for each crewman. Each man-meal package contains a 3.5-by-four-inch wet-wipe cleansing towel. Additionally, three packages of 12-by-12-inch dry towels are stowed beneath the command module pilot's couch. Each package contains seven towels. Also stowed under the command module pilot's couch are seven tissue dispensers containing 53 three-ply tissues each.

Solid body wastes are collected in Gemini-type plastic defecation bags which contain a germicide to prevent bacteria and gas formation. The bags are sealed after use and stowed in empty food containers for post-flight analysis.

Urine collection devices are provided for use while wearing either the pressure suit or the inflight coveralls. The urine is dumped overboard through the spacecraft urine dump valve in the CM and stored in the LM.

Medical Kit

The 5x5x8-inch medical accessory kit is stowed in a compartment on the spacecraft right side wall beside the lunar module pilot couch. The medical kit contains three motion sickness injectors, three pain suppression injectors, one two-ounce bottle first aid ointment, two one-ounce bottle eye drops, three nasal sprays, two compress bandages, 12 adhesive bandages, one oral thermometer, and four spare crew biomedical harnesses. Pills in the medical kit are 60 antibiotic, 12 nausea, 12 stimulant, 18 pain killer, 60 decongestant, 24 diarrhea, 72 aspirin and 21 sleeping. Additionally, a small medical kit containing four stimulant, eight diarrhea, two sleeping and four pain killer pills, 12 aspirin, one bottle eye drops and two compress bandages is stowed in the lunar module flight data file compartment.

Survival Gear

The survival kit is stowed in two rucksacks in the right hand forward equipment bay above the lunar module pilot.

Contents of rucksack No. 1 are: two combination survival lights, one desalter kit, three pair sunglasses, one radio beacon, one spare radio beacon battery and spacecraft connector cable, one knife in sheath, three water containers, and two containers of Sun lotion.

Rucksack No. 2: one three-man life raft with CO_2 inflater, one sea anchor, two sea dye markers, three sunbonnets, one mooring lanyard, three manlines, and two attach brackets.

The survival kit is designed to provide a 48-hour postlanding (water or land) survival capability for three crewmen between 40 degrees North and South latitudes.

Biomedical Inflight Monitoring

The Apollo 12 crew biomedical telemetry data received by the Manned Space Flight Network will be relayed for instantaneous display at Mission Control Center where heart rate and breathing rate data will be displayed on the flight surgeon's console. Heart rate and respiration rate average, range and deviation are computed and displayed on digital TV screens.

In addition, the instantaneous heart rate, real-time and delayed EKG and respiration are recorded on strip charts for each man.

Biomedical telemetry will be simultaneous from all crewmen while in the CSM, but selectable by a manual onboard switch in the LM.

Biomedical data observed by the flight surgeon and his team in the Life Support Systems Staff Support Room will be correlated with spacecraft and space suit environmental data displays.

Blood pressures are no longer telemetered as they were in the Mercury and Gemini programs. Oral temperature, however, can be measured onboard for diagnostic purposes and voiced down by the crew in case of inflight illness.

Training — The crewmen of Apollo 12 have spent more than five hours of formal crew training for each hour of the lunar-orbit mission's eight-day duration. More than 1,000 hours of training were in Apollo 12 crew training syllabus over and above the normal preparations for the mission-technical briefings and reviews, pilot meetings and study. As Apollo 9 backup crew, they had already received more than 1,500 hours of training.

The Apollo 12 crewmen also took part in spacecraft manufacturing checkouts at the North American Rockwell plant in Downey., Calif., at Grumman Aircraft Engineering Corp., Bethpage, N.Y., and in prelaunch testing at NASA Kennedy Space Center. Taking part in factory and launch area testing has provided the crew with thorough operational knowledge of the complex vehicle.

Highlights of specialized Apollo 12 crew training topics are:

* Detailed series of briefings on spacecraft systems, operation and modifications.

* Saturn launch vehicle briefings on countdown, range safety, flight dynamics, failure modes and abort conditions. The launch vehicle briefings were updated periodically.

* Apollo Guidance and Navigation system briefings at the Massachusetts Institute of Technology Instrumentation Laboratory.

* Briefings and continuous training on mission photographic objectives and use of camera equipment.

* Extensive pilot participation in reviews of all flight procedures for normal as well as emergency situations.

* Stowage reviews and practice in training sessions in the spacecraft, mockups and command module simulators allowed the crewmen to evaluate spacecraft stowage of crew-associated equipment.

* More than 400 hours of training per man in command module and lunar module simulators at MSC and KSC, including closed-loop simulations with flight controllers in the Mission Control Center. Other Apollo simulators at various locations were used extensively for specialized crew training.

* Entry corridor deceleration profiles at lunar-return conditions in the MSC Flight Acceleration Facility manned centrifuge.

* Lunar surface briefings and 1-g walk-throughs of lunar surface EVA operations covering lunar geology and microbiology and deployment of experiments in the Apollo Lunar Surface Experiment Package (ALSEP). Training in lunar surface EVA included practice sessions with lunar surface sample gathering tools and return containers, cameras, the erectable S-band antenna and the modular equipment stowage assembly (MESA) housed in the LM descent stage.

* Proficiency flights in the lunar landing training vehicle (LLTV) for the commander.

* Zero-g aircraft flights using command module and lunar module mockups for EVA and pressure suit doffing/donning practice and training.

* Underwater zero-g training in the MSC Water Immersion Facility using spacecraft mockups to further familiarize crew with all aspects of CSM-LM docking tunnel intravehicular transfer and EVA in pressurized suits.

* Water egress training conducted in indoor tanks as well as in the Gulf of Mexico, included uprighting from the Stable II position (apex down) to the Stable I position (apex up), egress onto rafts donning Biological Isolation Garments (BIGs), decontamination procedures and helicopter pickup.

* Launch pad egress training from mockups and from the actual spacecraft on the launch pad for possible emergencies such as fire, contaminants and power failures.

* The training covered use of Apollo spacecraft fire suppression equipment in the cockpit.

* Planetarium reviews at Morehead Planetarium, Chapel Hill, N.C., and at Griffith Planetarium, Los Angeles, Calif., of the celestial sphere with special emphasis on the 37 navigational stars used by the Apollo guidance computer.

CREW BIOGRAPHIES

NAME: Charles Conrad, Jr. (Commander, USN) Apollo 12 Commander NASA Astronaut

BIRTHPLACE AND DATE: Born on June 2, 1930, in Philadelphia, Pa.

PHYSICAL DESCRIPTION: Blond hair; blue eyes; height: 5 feet 6 1/2 inches; weight: 138 pounds.

EDUCATION: Attended primary and secondary schools in Haverford, Pa., and New Lebanon, New York; received a Bachelor of Science degree in Aeronautical Engineering from Princeton University in 1953 and an Honorary Master of Arts degree from Princeton in 1966.

MARITAL STATUS: Married to the former Jane DuBose of Uvalde, Texas, where her parents, Mr. and Mrs. W.O. DuBose, now reside.

CHILDREN:. Peter, December 24, 1954; Thomas, May 3, 1957; Andrew, April 30, 1959; Christopher, November 26, 1960.

OTHER ACTIVITIES: His hobbies include golf, swimming, and water skiing.

ORGANIZATIONS: Member of the American Institute of Aeronautics and Astronautics and the Society of Experimental Test Pilots.

SPECIAL HONORS: Awarded two Distinguished Flying Crosses, two NASA Exceptional Service Medals, and the Navy Astronaut Wings; recipient of Princeton's Distinguished Alumnus Award for 1965, the U.S. Jaycees' 10 Outstanding Young Men Award in 1965, and the American Astronautical Society Flight Achievement Award for 1966.

EXPERIENCE: Conrad entered the Navy following his graduation from Princeton University and became a naval aviator. He attended the Navy Test Pilot School at Patuxent River, Maryland, and upon completing that course of instruction was assigned as a project test pilot in the armaments test division there. He also served at Patuxent as a flight instructor and performance engineer at the Test Pilot School.

He has logged more than 4,000 hours flight time, with more than 3,000 hours in jet aircraft.

SALARY: $1,554.08 per month in military pay and allowances.

CURRENT ASSIGNMENT: Commander Conrad was selected as an astronaut by NASA in September 1962. In August 1965, he served as pilot on the 8-day Gemini 5 flight. He and command pilot Gordon Cooper were launched into orbit on August 21 and proceeded to establish a new space endurance record of 190 hours and 56 minutes. The flight, which lasted 120 revolutions and covered a total distance of 3,312,993 statute miles, was terminated on August 29, 1965. It was also on this flight that the United States took over the lead in man-hours in space.

On September 12, 1966, Conrad occupied the command pilot seat for the 3-day 44-revolution Gemini 11 mission. He executed orbital maneuvers to rendezvous and dock in less than one orbit with a previously launched Agena and controlled Gemini 11 through two periods of extravehicular activity performed by pilot Richard Gordon.

Other highlights of the flight included the retrieval of a nuclear emulsion experiment package during the first EVA; establishing a new world space altitude record of 850 statute miles; the successful completion of the first tethered station-keeping exercise, in which artificial gravity was produced; and the successful completion of the first fully automatic controlled reentry.

The flight was concluded on September 15, 1966, with the spacecraft landing in the Atlantic, 2½ miles from the prime recovery ship USS GUAM.

He served as backup commander for the Apollo 9 flight prior to his assignment as Apollo 12 commander.

Conrad has logged a total of 222 hours and 12 minutes of space flight in two missions.

NAME: Richard F. Gordon, Jr. (Commander, USN) Apollo 12 Command Module Pilot, NASA Astronaut

BIRTHPLACE AND DATE: Born October 5. 1929, in Seattle, Washington. His mother, Mrs. Angela Gordon, resides in Seattle.

PHYSICAL DESCRIPTION: Brown hair; hazel eyes; height: 5 feet 7 inches; weight: 150 pounds.

EDUCATION: Graduated from North Kitsap High School, Poulsbo, Washington; received a Bachelor of Science degree in Chemistry from the University of Washington in 1951.

MARITAL STATUS: Married to the former Barbara J. Field of Seattle, Washington. Her parents, Mr. and Mrs. Chester Field, reside in Freeland, Washington.

CHILDREN: Carleen, July 8 1954; Richard, October 6. 1955; Lawrence, December 18, 1957; Thomas, March 25, 1959; James, April 26, 1960; Diane, April 23, 1961.

OTHER ACTIVITIES: He enjoys water skiing, sailing, and golf.

ORGANIZATIONS: Member of the Society of Experimental Test Pilots.

SPECIAL HONORS: Awarded two Distinguished Flying Crosses, the NASA Exceptional Service Medal, and the Navy Astronaut Wings.

EXPERIENCE: Gordon, a Navy Commander, received his wings as a naval aviator in 1953. He then attended All-Weather Flight School and jet transitional training and was subsequently assigned to an all-weather fighter squadron at the Naval Air Station at Jacksonville, Fla.

In 1957, he attended the Navy's Test Pilot School at Patuxent River, Maryland, and served as a flight test pilot until 1960. During this tour of duty, he did flight test work on the F8U Crusader, F11F Tigercat, FJ Fury, and A4D Skyhawk and was the first project test pilot for the F4H Phantom II.

He served with Fighter Squadron 121 at the Miramar, Calif., Naval Air Station as a flight instructor in the F4H and participated in the introduction of that aircraft to the Atlantic and Pacific fleets. He was also flight safety officer, assistant operations officer, and ground training officer for Fighter Squadron 96 at Miramar.

Winner of the Bendix Trophy Race from Los Angeles to New York in May 1961, he established a new speed record of 869.74 miles per hour and a transcontinental speed record of 2 hours and 47 minutes. He was also a student at the U.S. Naval Postgraduate School at Monterey, California.

He has logged more than 4,038 hours flying time 3008 hours in jet aircraft.

SALARY: $1,633.28 per month in military pay and allowance.

CURRENT ASSIGNMENT: Commander Gordon was one of the third group of astronauts named by NASA in October 1963. He has since served as backup pilot for the Gemini 8 flight.

On September 12, 1966, he served as pilot for the 3-day 44-revolution Gemini 11 mission on which rendezvous with an Agena was achieved in less than one orbit. He executed docking maneuvers with the previously launched Agena and performed two periods of extravehicular activity which included attaching a tether to the Agena and retrieving a nuclear emulsion experiment package. Other highlights of the flight included the successful completion of the first tethered station-keeping exercises, the establishment of a new record setting altitude of 850 miles, and the first closed-loop controlled reentry.

The flight was concluded on September 15, 1966, with the spacecraft landing in the Atlantic, 2½ miles from the prime recovery ship, USS GUAM.

He served as Apollo 9 backup command module pilot prior to being named Apollo 12 command module pilot. Gordon has logged 71 hours 17 minutes of space flight, two hours and 44 minutes of which were in EVA.

NAME: Alan L. Bean (Commander, USN) Apollo 12 Lunar Module Pilot, NASA Astronaut

BIRTHPLACE AND DATE: Born in Wheeler, Texas, on March 15, 1932. His parents, Mr. and Mrs. Arnold H. Bean, reside in his hometown Fort Worth, Texas.

PHYSICAL DESCRIPTION: Brown hair; hazel eyes; height: 5 feet 9½ inches; weight: 155 pounds.

EDUCATION: Graduated from Paschal High School in Fort Worth, Texas; received a Bachelor of Science degree in Aeronautical Engineering from the University of Texas in 1955.

MARITAL STATUS: Married to the former Sue Ragsdale of Dallas, Texas; her parents, Mr. and Mrs. Edward B. Ragsdale, are residents of that city.

CHILDREN: Clay A., December 18, 1955; Amy Sue, January 21, 1963.

OTHER ACTIVITIES: His hobbies are playing with his two children, surfing, painting, and handball; and he also enjoys swimming, diving, and gymnastics.

ORGANIZATIONS: Member of the Society of Experimental Test Pilots and Delta Kappa Epsilon.

EXPERIENCE: Bean, a Navy ROTC student at Texas, was commissioned upon graduation in 1955. Upon completing his flight training, he was assigned to Attack Squadron 44 at the Naval Air Station in Jacksonville, Florida, for four years. He then attended the Navy Test Pilot School at Patuxent River, Maryland. Upon graduation he was assigned as a test pilot at the Naval Air Test Center, Patuxent River, where he flew all types of naval aircraft (jet, propeller, and helicopter models) to evaluate their suitability for operational Navy use. Commander Bean participated in the initial trials of both the A5A and A4E jet attack airplanes. He attended the school of Aviation Safety at the University of Southern California and was next assigned to Attack Squadron 172 at Cecil Field, Florida, as an A-4 light jet attack pilot.

During his career, he has flown 27 aircraft and logged more than 3,775 hours flying time, including 3,212 hours in jet aircraft.

SALARY: $1,071.08 per month in military pay and allowance.

CURRENT ASSIGNMENT: Commander Bean was one of the third group of astronauts selected by NASA in October 1963. He served as backup command pilot for the Gemini 10 mission and as the backup lunar module pilot for Apollo 9 prior to being named to the Apollo 12 crew as Lunar Module Pilot.

LAUNCH COMPLEX 39

Launch Complex 39 facilities at the Kennedy Space Center were planned and built specifically for the Apollo Saturn V, the space vehicle being used in the United States' manned lunar exploration program.

Complex 39 introduced the mobile concept of launch operations in which the space vehicle is thoroughly checked out in an enclosed building before it is moved to the launch pad for final preparations. This affords greater protection from the elements and permits a high launch rate since pad time is minimal.

Saturn V stages are shipped to the Kennedy Space Center by ocean-going vessels and specially designed aircraft. Apollo spacecraft modules are transported by air and first taken to the Manned Spacecraft Operations Building in the Industrial Area south of Complex 39 for preliminary checkout, altitude chamber testing and assembly.

Apollo 12 is the sixth Saturn V/Apollo space vehicle to be launched from Complex 39's Pad A. one of two octagonal launch pads which are 3,000 feet across. The major components of Complex 39 include:

1. The Vehicle Assembly Building, heart of the complex, is where the 363-foot-tall space vehicle is assembled and tested. It contains 129.5 million cubic feet of space, covers eight acres, is 716 feet long and 518 feet wide. Its high bay area, 525 feet high, contains four assembly and checkout bays and its low bay area - 210 feet high,442 feet wide and 274 feet long - contains eight stage-preparation and checkout cells. There are 141 lifting devices in the building, ranging from one-ton hoists to two 250-ton high lift bridge cranes.

2. The Launch Control Center, a four-story structure adjacent and to the south of the Vehicle Assembly Building is a radical departure from the dome-shaped, "hardened" blockhouse at older launch sites. The Launch Control Center is the electronic "brain" of Complex 39 and was used for checkout and test operations while Apollo 12 was being assembled inside the Vehicle Assembly Building high bay. Three of the four firing rooms contain identical sets of control and monitoring equipment so that launch of one vehicle and checkout of others may continue simultaneously. Each firing room is associated with a ground computer facility to provide data links with the launch vehicle on its mobile launcher at the pad or inside the Vehicle Assembly Building.

3. The Mobile Launcher, 445 feet tall and weighing 12 million pounds, is a transportable launch base and umbilical tower for the space vehicle.

4. The Transporters, used to move mobile launchers into the Vehicle Assembly Building and then with their space vehicles - to the launch pad, weigh six million pounds and are among the largest tracked vehicles known. The Transporters - there are two - are 131 feet long and 114 feet wide. Powered by electric motors driven by two 2,750-horsepower diesel engines, the vehicles move on four double-tracked crawlers, each 10 feet high and 40 feet long. Maximum speed is about one-mile-per-hour loaded and two miles-per-hour unloaded. The three and one-half mile trip to Pad A with a mobile launcher and space vehicle takes approximately seven hours. Apollo 12 rollout to the pad occurred on September 8, 1969.

5. The Crawlerway is the roadway for the transporter and is 131 feet wide divided by a median strip. This is the approximate width of an eight-lane turnpike and the roadbed is designed to accommodate a combined weight of more than 18 million pounds.

6. The Mobile Service Structure is a 402-foot-tall, 9.8 million pound tower used to service the Apollo space vehicle at the pad. Moved into place about the Saturn V/Apollo space vehicle and its mobile launcher by a transporter, it contains five work platforms and provides 360-degree platform access to the vehicle being prepared for launch. It is removed to a parking area about 11 hours before launch.

7. A Water Deluge System will provide about a million gallons of industrial water for cooling and fire prevention during the launch of Apollo 12. The water is used to cool the mobile launcher, the flame trench and the flame deflector above which the mobile launcher is positioned.

8. The Flame Deflector is an "A"-shaped, 1.3 million pound structure moved into the flame trench beneath the launcher prior to launch. It is covered with a refractory material designed to withstand the launch environment. The flame trench itself is 58 feet wide and approximately six feet above mean sea level at the base.

9. The Pad Areas - A and B - are octagonal in shape and have center hardstands constructed of heavily reinforced concrete. The top of Pad A stands about 48 feet above sea level. Saturn V propellants - liquid oxygen, liquid hydrogen and RP-1, the latter a high grade kerosene - are stored in large tanks spaced near the pad perimeter and carried by pipelines from the tanks to the pad, up the mobile launcher and into the launch vehicle propellant tanks. Also located in the pad area are pneumatic, high pressure gas, electrical, and industrial water support facilities. Pad B, used for the launch of Apollo 10, is located 8,700 feet north of Pad A.

MISSION CONTROL CENTER

The Mission Control Center at the Manned Spacecraft Center, Houston, is the focal point for Apollo flight control activities. The center receives tracking and telemetry data from the Manned Space Flight Network which in turn is processed by the MCC Real-Time Computer Complex for display to flight controllers in the Mission Operations Control Room (MOCR) and adjacent staff support rooms.

Console positions in the two identical MOCRs in Mission Control Center fall into three basic operations groups: mission command and control, systems operations, and flight dynamics.

Positions in the command and control group are:

* Mission Director — responsible for overall mission conduct.

* Flight Operations Director — represents MSC management.

* Flight Director — responsible for operational decisions and actions in the MOCR.

* Assistant Flight Director — assists flight director and acts in his absence.

* Flight Activities Officer — develops and coordinates flight plan.

* Department of Defense Representative — coordinates and directs DOD mission support.

* Network Controller — responsible to FD for Manned Space Flight Network status and troubleshooting; MCC equipment operation.

* Surgeon — monitors crew medical condition and informs FD of any medical situation affecting mission.

* Spacecraft Communicator (Capcom) — serves as voice contact with flight crew.

* Experiments Officer — coordinates operation and control of onboard flight experiments.

* Public Affairs Officer — reports mission progress to public through commentary and relay of live air-to-ground transmissions.

Systems Operations Group:

* Environmental, Electrical and Communications Engineer (EECOM) - monitors and troubleshoots command/service module environmental, electrical, and sequential systems.

* Guidance, Navigation and Control Engineer (GNC) - monitors and troubleshoots CSM guidance, navigation, control, and propulsion systems.

* LM Environmental and Electrical Engineer (TELCOM) -LM counterpart to EECOM.

* LM Guidance, Navigation and Control Engineer (Control)-LM counterpart to GNC.

* Booster Systems Engineer (three positions) — responsible for monitoring launch vehicle performance and for sending function commands.

* Apollo Communications Engineer (ACE) and Operations and Procedures Officer (O&P) — share responsibility for monitoring and troubleshooting spacecraft and lunar surface communication systems and for coordinating MCC procedures with other NASA centers and the network.

Flight Dynamics Group:

* Flight Dynamics Officer (FIDO) — monitors powered flight events and plans spacecraft maneuvers.

* Retrofire Officer (Retro) — responsible for planning deorbit maneuvers in Earth orbit and entry calculations on lunar return trajectories.

* Guidance officer (Guido) — responsible for monitoring and updating CSM and LM guidance systems and for monitoring systems performance during powered flight. Each MOCR operations group has a staff support room on the same floor in which detailed monitoring and analysis is conducted. Other supporting MCC areas include the spaceflight Meteorological Room, the Space Environment (radiation) Console, Spacecraft Planning and Analysis (SPAN) Room for detailed spacecraft performance analysis, Recovery Operations Control Room and the Apollo Lunar Surface Experiment Package Support Room.

Located on the first floor of the MCC are the communications, command, and telemetry system (CCATS) for processing incoming data from the tracking network, and the real-time computer complex (RTCC) which converts flight data into displays useable to MOCR flight controllers.

MANNED SPACE FLIGHT NETWORK

The Manned Space Flight Network (MSFN) is a worldwide system that provides reliable, continuous, and instantaneous communications with the astronauts, launch vehicle, and spacecraft from liftoff to splashdown. The MSFN also will link between Earth and the Apollo experiments left on the lunar surface by the Apollo 12 crew.

The worldwide tracking network is maintained and operated by the NASA Goddard Space Flight Center (Greenbelt, Md.), under the direction of NASA's Office of Tracking and Data Acquisition. In the Manned Space Flight Network Operations Center (MSFNOC) at Goddard, the Network Director and his team of Operations Managers, with the assistance of a Network Support Team, keep the entire complex tuned for the mission support. Should Houston's mission control center be seriously impaired for an extended time, the Goddard Center becomes an emergency mission control center.

The MSFN employs 13 ground tracking stations equipped with 30 and 85 foot antennas, an instrumented tracking ship, and four instrumented aircraft. For Apollo 12, the network will be augmented by the 210-foot antenna systems at Goldstone, Calif. and at Parkes, Australia, (Australian Commonwealth Scientific and Industrial Research Organization).

NASA Communications Network (NASCOM)

The tracking network is linked together by the NASA communications network. All information flows to and from MCC Houston and the Apollo spacecraft over this communications system.

The NASCOM consists of almost three million circuit miles of diversely routed communications channels. It uses satellites, submarine cables, land lines, microwave systems, and high frequency radio facilities for access links.

NASCOM control center is located at Goddard. Regional communication switching centers are located in London, Madrid, Canberra, Australia, Honolulu, and Guam.

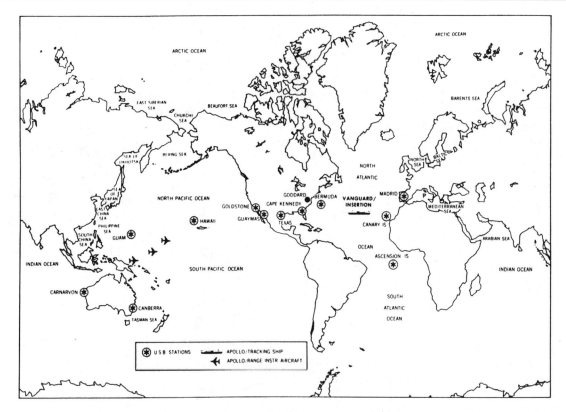

MANNED SPACE FLIGHT TRACKING NETWORK

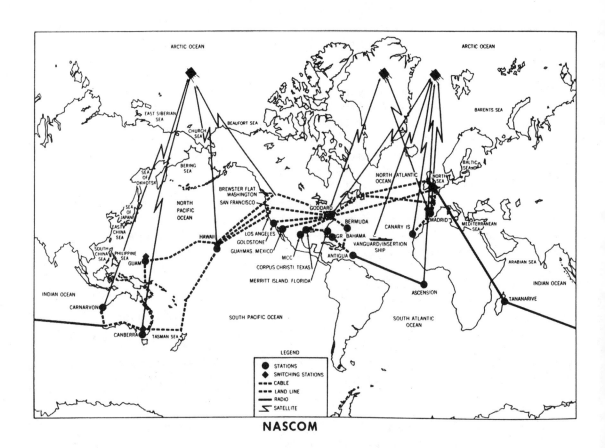

NASCOM

Three Intelsat communications satellites will be used for Apollo 12. One satellite over the Atlantic will link Goddard with stations at Madrid, Canary Islands, Ascension and the Vanguard tracking ship. Another Atlantic satellite will provide a direct link between Madrid and Goddard for TV signals received from the Apollo 12. The third satellite over the mid-Pacific will link the two Carnarvon and Canberra Australia and Hawaii with Goddard through a ground station at Brewster Flats, Wash.

At Goddard, NASCOM switching computers simultaneously send the voice signals directly to the Houston flight controllers and the tracking and telemetry data to computer processing complexes at Houston and Goddard. The Goddard Real Time Computing Complex verifies performance of the tracking network and uses the collected tracking data to drive displays in the Goddard Operations Control Center.

Establishing the Link — The Merritt Island tracking station monitors prelaunch test, the terminal countdown, and the first minutes of launch.

An Apollo instrumentation ship (USNS VANGUARD) fills the gaps beyond the range of land tracking stations. For Apollo 12, this ship will be stationed in the Atlantic to cover the insertion into Earth orbit. Apollo instrumented aircraft provide communications support to the land tracking stations during translunar injection and reentry and cover a selected abort area in the event of "no-go" decision after insertion into Earth orbit.

Lunar Bound — Approximately one hour after the spacecraft has been injected into its translunar trajectory (some 10,000 miles from the Earth), three prime tracking stations spaced nearly equidistant around the Earth will take over tracking and communicating with Apollo.

The prime stations are located at Goldstone, Calif.; Madrid, Spain; and Canberra, Australia. Each station has a dual system for use when tracking the command module in lunar orbit and the lunar module in separate flight paths or at rest on the Moon.

The Return Trip — To make an accurate reentry, data from the tracking stations are fed into the MCC computers to develop necessary information for the Apollo 12 crew.

Appropriate MSFN stations, including the aircraft in the Pacific, are on hand to provide support during the reentry. Through the journey to the Moon and return, television will be received from the spacecraft at the three 85-foot antennas around the world. In addition, the 210-foot antennas in California and Australia will be used to augment the television coverage while the Apollo 12 is near and on the Moon. For black and white TV, scan converters at the stations permit immediate transmission of commercial quality TV via NASCOM to Houston where it will be released to U.S. TV networks.

The black and white TV can be released simultaneously in Europe and the Far East through the MSFN stations in Spain and Australia.

If color TV is used, the signal will be converted to commercial quality at the MSC Houston.

Network Computers

At fraction-of-a-second intervals, the network's digital data processing systems, with NASA's Manned Spacecraft Center as the focal point, "talk" to each other or to the spacecraft. High-speed computers at the remote site (tracking ship included) relay commands or "up-link" data on such matters as control of cabin pressure, orbital guidance commands, or "go-no-go" indications to perform certain functions.

When information originates from Houston, the computers refer to their preprogrammed information for validity before transmitting the required data to the spacecraft.

Such "up-link" information is communicated at a rate of about 1,200 bits-per-second. Communication between remote ground sites, via high-speed communications links, occurs at about the same rate. Houston reads information, two channels at a time, from these ground sites at 2,400 bits-per-second.

The computer systems perform many other functions, including:

Assuring the quality of the transmission lines by continually exercising data paths;

Verifying accuracy of the messages.

Constantly updating the flight status.

For "down-link" data, sensors built into the spacecraft continually sample cabin temperature, pressure, physical information on the astronauts such as heartbeat and respiration. These data are transmitted to the ground stations at 51.2 kilobits (12,800 decimal digits) per second.

At MCC the computers:

Detect and select changes or deviations, compare with their stored programs, and indicate the problem areas or pertinent data to the flight controllers;

Provide displays to mission personnel;

Assemble output data in proper formats;

Log data on magnetic tape for reply for the flight controllers.

The Apollo Ship Vanguard

The USNS Vanguard will perform tracking, telemetry, and communication functions for the launch phase and Earth orbit insertion. Vanguard will be stationed about 1,000 miles southeast of Bermuda (28 degrees N., 49 degrees W.).

Apollo Range Instrumentation Aircraft (ARIA)

During Apollo 12 TLI maneuver, two ARIA will record telemetry data from Apollo and relay voice communication between the astronauts and the Mission Control Center at Houston. The ARIA will be located between Australia and Hawaii.

For reentry, two ARIA will be deployed to the landing area to relay communications between Apollo and Mission Control at Houston and provide position information on the spacecraft after the blackout phase of reentry has passed.

The total ARIA fleet for Apollo missions consists of four EC-135A (Boeing 707) jets with 7-foot parabolic antennas installed in the nose section.

Lunar Receiving Laboratory (LRL)

The final phase of the back contamination program is completed in the MSC Lunar Receiving Laboratory. The crew and spacecraft are quarantined for a minimum of 21 days after completion of lunar EVA operations and are released based upon the completion of prescribed test requirements and results. The lunar sample will be quarantined for a period of 50 to 80 days depending upon results of extensive biological tests.

The LRL serves four basic purposes:

* Quarantine of crew and spacecraft, the containment of lunar and lunar-exposed materials, and quarantine

testing to search for adverse effects of lunar material upon terrestrial life.

* The preservation and protection of the lunar samples.

* The performance of time critical investigations.

* The preliminary examination of returned samples to assist in an intelligent distribution of samples to principal investigators.

The LRL has the only vacuum system in the world with space gloves operated by a man leading directly into a vacuum chamber at pressures of 10^{-7} torr. (mm Hg) (or one 10 billionth of an atmosphere). It has a low level counting facility, whose background count is an order of magnitude better than other known counters. Additionally, it is a facility that can handle a large variety of biological specimens inside Class III biological cabinets designed to contain extremely hazardous pathogenic material.

The LRL covers 83,000 square feet of floor space and includes a Crew Reception Area (CRA), Vacuum Laboratory, Sample Laboratories (Physical and Bio-Science) and an administrative and support area. Special building systems are employed to maintain air flow into sample handling areas and the CRA, to sterilize liquid waste, and to incinerate contaminated air from the primary containment systems.

The biomedical laboratories provide for quarantine tests to determine the effect of lunar samples on terrestrial life. These tests are designed to provide data upon which to base the decision to release lunar material from quarantine.

Among the tests:

a. Germ-free mice will be exposed to lunar material and observed continuously for 21 days for any abnormal changes. Periodically, groups will be sacrificed for pathologic observation.

b. Lunar material will be applied to 12 different culture media and maintained under several environmental conditions. The media will be observed for bacterial or fungal growth. Detailed inventories of the microbial flora of the spacecraft and crew have been maintained so that any living material found in the sample testing can be compared against this list of potential contaminants taken to the Moon by the crew or spacecraft.

c. Six types of human and animal tissue culture cell lines will be maintained in the laboratory and together with embryonated eggs are exposed to the lunar material. Based on cellular and/or other changes, the presence of viral material can be established so that special tests can be conducted to identify and isolate the type of virus present.

d. Thirty-three species of plants and seedlings will be exposed to lunar material. Seed germination, growth of plant cells or the health of seedlings are then observed, and histological, microbiological and biochemical techniques are used to determine the cause of any suspected abnormality.

e. A number of lower animals will be exposed to lunar material, including fish, birds, oysters, shrimp, cockroaches, houseflies, planaria, paramecia and euglena. If abnormalities are noted, further tests will be conducted to determine if the condition is transmissible from one group to another.

The crew reception area provides biological containment for the flight crew and 12 support personnel. The nominal occupancy is about 14 days but the facility is designed and equipped to operate for considerably longer.

Sterilization and Release of the Spacecraft

Post-flight testing and inspection of the spacecraft is presently limited to investigation of anomalies which happened during the flight. Generally, this entails some specific testing of the spacecraft and removal of certain components of systems for further analysis. The timing of post-flight testing is important so that corrective action may be taken for subsequent flights.

The schedule calls for the spacecraft to be returned to port where a team will deactivate pyrotechnics, and flush and drain fluid systems (except water). This operation will be confined to the exterior of the spacecraft. The spacecraft will then be flown to the LRL and placed in a special room for storage, sterilization, and post-flight checkout.

The Interagency Committee on Back Contamination (ICBC) functions to assist NASA in the program to prevent contamination of the Earth from lunar materials. The ICBC met Oct. 30 in Atlanta to review the Apollo 12 contamination control procedures.

LUNAR RECEIVING LABORATORY TENTATIVE SCHEDULE

Nov. 19	Activate secondary barrier; support people enter Crew Reception Area and Central Status Station manned; LRL on mission status.
Nov. 24	Command module landing, recovery.
Nov. 25	First sample return container (SRC) arrives.
Nov. 26	First SRC opened in vacuum lab, second SRC arrives; film, tapes, LM tape recorder begin decontamination; second SRC opened in Bioprep lab.
Nov. 27	First sample to Radiation Counting Laboratory.
Nov. 29	Core tube moves from vacuum lab to Physical-Chemical Lab.
Nov. 30	MQF arrives; contingency sample goes to Physical-Chemical Lab; rock description begun in vacuum lab.
Dec. 1	Biosample rocks move from vacuum lab to Bioprep Lab; core tube prepared for biosample.
Dec. 3	Spacecraft arrives.
Dec. 4	Biosample compounded, thin-section chips sterilized out to Thin-Section Lab, remaining samples from Bioprep Lab canned.
Dec. 6	Thin-section preparation complete, biosample prep complete, transfer to Physical-Chemical Lab complete, Bioprep Lab cleanup complete.
Dec. 8	Biological protocols, Physical-Chemical Lab rock description begin.
Dec. 10	Crew released from CRA.
Dec. 12	Spacecraft released.
Dec. 20	Rock description complete, Preliminary Examination Team data from Radiation Counting Lab and Gas Analysis Lab complete.
Dec. 22	PET data write-up and sample catalog preparation begin.
Dec. 24	Data summary for Lunar Sample Analysis Planning Team (LSAPT) complete.
Dec. 26	LSAPT arrives.
Dec. 27	LSAPT briefed on PET data, sample packaging begins.

Dec. 31	Sample distribution plan complete, first batch monopole samples canned.
Jan. 2	Monopole experiment begins.
Jan. 5	Apollo 11 (sic) principal investigator conference begins.
Jan. 7	Sample distribution plan approved, sample release, sample catalog complete.
Jan. 8	Initial release of Apollo 12 samples.
Jan. 12	Spacecraft equipment released.
Jan. 16	Apollo 12 mission critique.

CONTAMINATION CONTROL PROGRAM

In 1966 an Interagency Committee on Back Contamination (ICBC) was established to assist NASA in developing a program to prevent contamination of the Earth from lunar materials following manned lunar exploration and to review and approve plans and procedures to prevent back contamination. Committee membership includes representatives from Public Health Service, Department of Agriculture, Department of the Interior, NASA, and the National Academy of Sciences.

The Apollo Back Contamination Program can be divided into three phases. The first phase covers procedures which are followed by the crew while in flight to reduce and, if possible, eliminate the return of lunar surface contaminants in the command module.

The second phase includes recovery, isolation, and transport of the crew, spacecraft, and lunar samples to the Manned Spacecraft Center. The third phase encompasses quarantine operations and preliminary sample analysis in the Lunar Receiving Laboratory.

A primary step in preventing back contamination is careful attention to spacecraft cleanliness following lunar surface operations. This includes use of special cleaning equipment, stowage provisions for lunar-exposed equipment, and crew procedures for proper "housekeeping."

Prior to reentering the LM after lunar surface exploration, the crewmen brush lunar surface dust or dirt from the space suit using the suit gloves. They will scrape their over-boots on the LM footpad and while ascending the LM ladder, dislodge any clinging particles by a kicking action.

After entering and pressurizing the LM cabin, the crew doff their portable life support system, oxygen purge system, lunar boots, EVA gloves, etc.

Following LM rendezvous and docking with the CM, the CM tunnel will be pressurized and checks made to insure that an adequate pressurized seal has been made. During the period, the LM, space suits, and lunar surface equipment will be vacuumed.

The lunar module cabin atmosphere will be circulated through the environmental control system suit circuit lithium hydroxide (LiOH) canister to filter particles from the atmosphere. A minimum of five hours weightless operation and filtering will essentially eliminate the original airborne particles.

The CM pilot will transfer lunar surface equipment stowage bags into the LM one at a time. The equipment transferred will be bagged before being transferred. The only equipment which will not be bagged at this time are the crewmen's space suits and flight logs.

<u>Command Module Operations</u> — Through the use of operational and housekeeping procedures the command module cabin will be purged of lunar surface and/or other particulate contamination prior to Earth reentry. These procedures start while the LM is docked with the CM and continue through reentry into the Earth's atmosphere.

During subsequent lunar orbital flight and the transearth phase, the command module atmosphere will be continually filtered through the environmental control system lithium hydroxide canister. This will remove essentially all airborne dust particles. After about 96 hours operation essentially none of the original contaminates will remain.

<u>Lunar Mission Recovery Operations</u>

Following landing and the attachment of the flotation collar to the command module, the swimmer in a biological isolation garment (BIG) will open the spacecraft hatch, pass three BIGs into the spacecraft, and close the hatch.

The crew will don the BIG's and then egress into a life raft. The hatch will be closed immediately after egress. Tests have shown that the crew can don their BIG's in less than 5 minutes under ideal sea conditions. The spacecraft and crew will be decontaminated by the swimmer using a liquid agent.

Crew retrieval will be accomplished by helicopter to the carrier and subsequent crew transfer to the Mobile Quarantine Facility. The spacecraft will be retrieved by the aircraft carrier and isolated.

SCHEDULE FOR TRANSPORT OF SAMPLES, SPACECRAFT AND CREW

<u>Samples</u>

The first Apollo 12 lunar sample return container will be flown by carrier on-board delivery (COD) aircraft from the deck of the USS Hornet to Samoa, from where it will be flown by USAF C-141 to Ellington AFB about 21 hours after spacecraft touchdown.

The second sample container will leave the Hornet by COD aircraft about 13 hours after splashdown, transfer at Samoa to an ARIA aircraft, and after a refueling stop in Hawaii, arrive at Ellington AFB about 30 hours after spacecraft splashdown. Both sample return flights will include medical supplies, spacecraft onboard film and other equipment. The shipments will be moved by auto to the Lunar Receiving Laboratory.

<u>Spacecraft</u>

The spacecraft is scheduled to be brought aboard the Hornet about two hours after crew recovery. About four days, 19 hours after recovery the ship is expected to arrive in Hawaii. The spacecraft will be deactivated in Hawaii between 115 and 166 hours after recovery. At 166 hours it is scheduled to be loaded on a C-133B for return to Ellington AFB. Estimated time of arrival at the LRL is on Dec. 2, 198 hours after recovery.

<u>Crew</u>

The flight crew is expected to enter the Mobile Quarantine Facility (MQF) on the recovery ship about 90 minutes after splashdown. The ship is expected to arrive in Hawaii at recovery plus 115 hours and the Mobile Quarantine Facility will be transferred to a C-141 aircraft at Pearl Harbor at recovery plus 117 hours. The aircraft will land at Ellington AFB at recovery plus 123 hours and the MQF will arrive at the LRL about two hours later Nov. 29.

APOLLO PROGRAM MANAGEMENT

The Apollo Program is the responsibility of the Office of Manned Space Flight (OMSF), National Aeronautics and Space Administration, Washington, D. C. Dr. George E. Mueller is Associate Administrator for Manned

Space Flight.

NASA Manned Spacecraft Center (MSC), Houston, is responsible for development of the Apollo spacecraft, flight crew training, and flight control. Dr. Robert R. Gilruth is Center Director.

NASA Marshall Space Flight Center (MSFC), Huntsville, Ala., is responsible for development of the Saturn launch vehicles. Dr. Wernher von Braun is Center Director

NASA John F. Kennedy Space Center (KSC), Fla., is responsible for Apollo/Saturn launch operations. Dr. Kurt H. Debus is Center Director.

The NASA Office of Tracking and Data Acquisition (OTDA) directs the program of tracking and data flow on Apollo. Gerald M. Truszynski is Associate Administrator for Tracking and Data Acquisition.

NASA Goddard Space Flight Center (GSFC), Greenbelt, Md., manages the Manned Space Flight Network and Communications Network. Dr. John F. Clark is Center Director.

The Department of Defense is supporting NASA in Apollo 12 during launch, tracking and recovery operations. The Air Force Eastern Test Range is responsible for range activities during launch and down-range tracking. Recovery operations include the use of recovery ships and Navy and Air Force aircraft.

Apollo/Saturn Officials

NASA Headquarters

Dr. Rocco A. Petrone	Apollo Program Director, OMSF
Capt. Chester M. Lee, (USN, Ret.)	Apollo Mission Director, OMSF
Col. Thomas H. McMullen (USAF)	Assistant Mission Director, OMSF
Maj. Gen. James W. Humphreys, Jr.	Director of Space Medicine, OMSF
Norman Pozinsky	Director, Network Support Implementation Div., OTDA

Manned Spacecraft Center

Col. James A. McDivitt, (USAF)	Manager, Apollo Spacecraft Program
Kenneth S. Kleinknecht	Manager, Command and Service Modules
Owen G. Morris	Manager, Lunar Module (Acting)
Donald K. Slayton	Director of Flight Crew Operations
Christopher C. Kraft,, Jr.	Director of Flight Operations
Gerald Griffin	Flight Director
Glynn S. Lunney	Flight Director
Clifford E. Charlesworth	Flight Director
M. P. Frank	Flight Director
Charles A. Berry	Director of Medical Research and Operations

Marshall Space Flight Center

Lee B. James	Director, Program Management
Dr. F. A. Speer	Manager, Mission Operations Office
Roy E. Godfrey	Manager, Saturn Program Office
Matthew W. Urlaub.	Manager, S-IC Stage, Saturn Program Office
W. F. LaHatte	Manager, S-II Stage, Saturn Program Office
James C. McCulloch	Manager, S-IVB Stage, Saturn Program Office
Frederich Duerr	Manager, Instrument Unit, Saturn Program Office
William D. Brown	Manager, Engine Program Office

Kennedy Space Center

Walter J. Kapryan   Director, Launch Operations
Edward R. Mathews   Manager, Apollo Program Office
Dr. Hans F. Gruene   Director, Launch Vehicle Operations
John J. Williams   Director, Spacecraft Operations
Paul C. Donnelly   Launch Operations Manager

Goddard Space Flight Center

Ozro M. Covington   Director of Manned Flight Support
William P. Varson   Chief, Manned Flight Planning & Analysis Division
H. William Wood   Chief, Manned Flight Operations Division
Tecwyn Roberts   Chief, Manned Flight Engineering Division
L. R. Stelter   Chief, NASA Communications Div.

Department of Defense

Maj. Gen. David M. Jones, (USAF)   DOD Manager of Manned Space Flight Support Operations,
                                            Commander of USAF Eastern Test Range
Rear Adm. Donald C. Davis,(USN)   Commander of Combined Task Force
                                            130, Pacific Recovery Area

Rear Adm. Philip S. McManus (USN)   Commander of Combined Task Force 140,
                                            Atlantic Recovery Area

Col. Royce G. Olson (USAF)   Director of DOD Manned Space Flight Office
Brig. Gen. Allison C. Brooks, (USAF)   Commander Aerospace Rescue and Recovery Service

Major Apollo/Saturn V ContractorsContractor   Item

Bellcomm   Apollo Systems Engineering
Washington, D. C.

The Boeing Co.   Technical Integration and
Washington, D. C.   Evaluation

General Electric-   Apollo Systems, Apollo Checkout, and Quality and Reliability
Daytona Beach, Fla

North American Rockwell Corp.   Command and Service Modules
Space Div., Downey, Calif.

Grumman Aircraft Engineering   Lunar Module
Corp., Bethpage, N.Y.

Massachusetts Institute of   Guidance & Navigation
Technology, Cambridge, Mass.   (Technical Management)

General Motors Corp., AC   Guidance & Navigation
Electronics Div, Milwaukee, Wis.   (Manufacturing)

TRW Inc.   Trajectory Analysis
Systems Group   LM Descent Engine
Redondo Beach, Calif.   LM Abort Guidance System

Avco Corp., Space Systems   Heat Shield Ablative Material
Div., Lowell, Mass.

North American Rockwell Corp.   J-2 Engines, F-1 Engines
Rocketdyne Div.Canoga Park, Calif.

The Boeing Co. New Orleans,	First Stage (SIC) of Saturn V Launch Vehicles, Saturn V Systems Engineering and Integration, Ground Support Equipment
North American Rockwell Corp. Space Div. Seal Beach, Calif.	Development and Production of Saturn V Second Stage (S-II)
McDonnell Douglas Astronautics Co., Huntington Beach, Calif.	Development and Production of Saturn V Third Stage (S-IVB)
International Business Machines Federal Systems Div.Huntsville, Ala.	Instrument Unit
Bendix Corp. Navigation and Control Div. Teterboro., N.J.	Guidance Components for Instrument Unit (Including ST-124M Stabilized Platform)
Federal Electric Corp.	Communications and Instrumentation Support, KSC
Bendix Field Engineering Corp.	Launch Operations/Complex Support, KSC
Catalytic-Dow	Facilities Engineering and Modifications, KSC
Hamilton Standard Division	Portable Life Support System;
United Aircraft Corp. Windsor Locks, Conn.	LM ECS
ILC Industries Dover, Del.	Space Suits
Radio Corp. of America Van Nuys, Calif.	110A Computer - Saturn Checkout
Sanders Associates Nashua, N.H.	Operational Display Systems Saturn
Brown Engineering Huntsville, Ala.	Discrete Controls
Reynolds, Smith and Hill Jacksonville, Fla.	Engineering Design of Mobile Launchers
Ingalls Iron Works Birmingham, Ala.	Mobile Launchers (ML) (Structural Work)
Smith/Ernst (Joint Venture) Tampa, Fla. Washington, D. C.	Electrical Mechanical Portion of MLs
Power Shovel, Inc. Marion, Ohio	Transporter
Hayes International Birmingham, Ala.	Mobile Launcher Service Arms
Bendix Aerospace Systems Ann Arbor, Mich.	Apollo Lunar Surface Experiments Package (ALSEP)
Aerojet-Gen. Corp. El Monte. Calif.	Service Propulsion System Engine

MISSION OPERATION REPORT APOLLO 12 (AS-507)

MISSION OPERATION REPORT

APOLLO 12 (AS-507)

OFFICE OF MANNED SPACE FLIGHT
PREPARED BY: APOLLO PROGRAM OFFICE - MAO

Report No. M-932-69-12

FOREWORD

MISSION OPERATION REPORTS are published expressly for the use of NASA Senior Management, as required by the Administrator in NASA Instruction 6-2-10, dated 15 August 1963. The purpose of these reports is to provide NASA Senior Management with timely, complete, and definitive information on flight mission plans, and to establish official mission objectives which provide the basis for assessment of mission accomplishment.

Initial reports are prepared and issued for each flight project just prior to launch. Following launch, updating reports for each mission are issued to keep General Management currently informed of definitive mission results as provided in NASA Instruction 6-2-10.

Primary distribution of these reports is intended for personnel having program/project management responsibilities which sometimes results in a highly technical orientation. The Office of Public Affairs publishes a comprehensive series of pre launch and post launch reports on NASA flight missions which are available for dissemination to the Press.

APOLLO MISSION OPERATION REPORTS are published in two volumes: the MISSION OPERATION REPORT (MOR); and the MISSION OPERATION REPORT, APOLLO SUPPLEMENT. This format was designed to provide a mission-oriented document in the MOR, with supporting equipment and facility description in the MOR, APOLLO SUPPLEMENT. The MOR, APOLLO SUPPLEMENT is a program-oriented reference document with a broad technical description of the space vehicle and associated equipment, the launch complex, and mission control and support facilities.

Pre launch
Mission Operation Report
No. M-932-69-12

TO: A/Administrator 5 November 1969

FROM: MA/Apollo Program Director

SUBJECT: Apollo 12 Mission (AS-507)

No earlier than 14 November 1969, we plan to launch Apollo 12 on the second lunar landing mission. This will be the fifth manned Saturn V flight, the sixth flight of a manned Apollo Command/Service Module, and the fourth flight of a manned Lunar Module.

Apollo 12 will be launched from Pad A of Launch Complex 39 at the Kennedy Space Center. Lunar touchdown is planned for Apollo Landing Site 7, located in the Ocean of Storms about 830 nautical miles west of the Apollo 11 landing site in the Sea of Tranquillity. Apollo Landing Site 7 includes the crater in which Surveyor III landed in April 1967. One of the primary objectives of this mission is to develop techniques for a point landing capability.

Primary objectives on the lunar surface include selenological inspection, survey, and sampling in a mare area; deployment and activation of an Apollo Lunar Surface Experiments Package; and development of man's capability to work in the lunar environment. Photographic records will be obtained and extravehicular activities will be televised.

Following the lunar surface phase of the mission, the crewmen will return to the Command/Service Module and remain in lunar orbit approximately 1 day to perform the remaining primary objective of obtaining extensive photography of candidate exploration sites for future missions.

The 10-day mission will be completed with landing in the Pacific Ocean. Recovery and transport of the crew, spacecraft, and lunar samples to the Lunar Receiving Laboratory at the Manned Spacecraft Center will be conducted under quarantine procedures that provide for biological isolation.

Rocco A. Petrone

APPROVAL:

George E. Mueller
Associate Administrator for
Manned Space Flight

PROGRAM DEVELOPMENT

Since the first Saturn flight, the Apollo Program has been developing toward a lunar landing and exploration of the lunar surface. Each successive flight has demonstrated the performance capabilities of the space vehicle, crew, and ground support and has verified operational techniques and procedures. The first Apollo flights, AS-201 through Apollo 6, were launch vehicle and spacecraft development flights. Apollo 7, the first manned flight, demonstrated Command/Service Module (CSM)/crew performance and CSM rendezvous capability. The Apollo 8 Mission carried CSM operations further by successfully demonstrating CSM operations and selected backup lunar landing mission activities in lunar orbit. Apollo 9 was an earth-orbital mission which demonstrated CSM/Lunar Module (LM) operations and LM/crew performance of selected lunar landing mission activities in earth orbit. The final developmental mission before the actual lunar landing was Apollo 10. It evaluated LM performance in the cislunar and lunar environment and duplicated the lunar landing mission profile as closely as possible without actually landing. The success of these missions finally culminated in the Apollo 11 Mission, the first manned lunar landing and return mission. The success of the Apollo 11 Mission verified the performance of the space vehicle and support systems and proved man's capability to accomplish a lunar mission enabling the Apollo Program to proceed with detailed exploration of the lunar surface. Figure 1 traces the Apollo flight mission development phases through the first lunar landing.

The final nine lunar exploration missions in the Apollo Program will be divided into two types of missions - H-series and J-series. The four H-series missions, Apollo 12 through Apollo 15, will be flown with standard Apollo hardware and will provide increased surface stay time with two extravehicular activity (EVA) periods, improved landing accuracy, development of CSM transport techniques, and will establish a seismic network. The last five missions, Apollo 16 through Apollo 20, will be J-series missions and will be flown with modified Apollo hardware designed to extend mission duration and lunar surface stay time, to increase landed payload and sample return, to extend lunar surface EVA operations and increase mobility, and to provide for scientific experiments and mapping to be accomplished in lunar orbit.

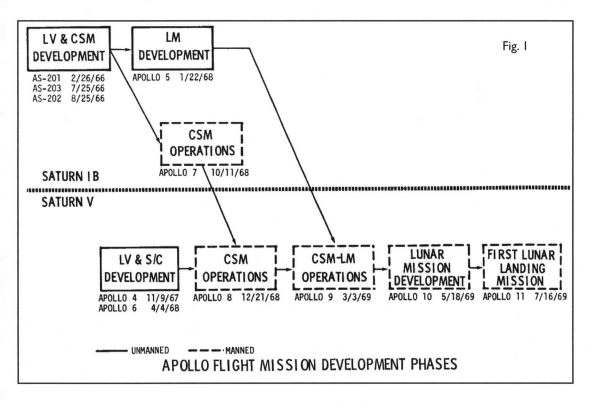

APOLLO FLIGHT MISSION DEVELOPMENT PHASES

NASA OMSF PRIMARY MISSION OBJECTIVES FOR APOLLO 12

PRIMARY OBJECTIVES

1. Perform selenological inspection, survey, and sampling in a mare area.
2. Deploy and activate an Apollo Lunar Surface Experiments Package (ALSEP).
3. Develop techniques for a point landing capability.
4. Develop man's capability to work in the lunar environment.
5. Obtain photographs of candidate exploration sites.

Rocco A. Petrone   George E. Mueller
Apollo Program Director   Associate Administrator for
                                                            Manned Space Flight
Date: 31 October 1969   Date: 3 November 1969

DETAILED OBJECTIVES AND EXPERIMENTS

PRINCIPAL DETAILED OBJECTIVES

1. Contingency Sample Collection.
2. Lunar Surface EVA Operations.
3. Apollo Lunar Surface Experiments Package (ALSEP) I Deployment and Activation.
4. Selected Sample Collection.
5. PLSS Recharge.
6. Lunar Field Geology (S-059).
7. Photography of Candidate Exploration Sites.
8. Lunar Surface Characteristics.
9. Lunar Environment Visibility.
10. Landed LM Location.
11. Selenodetic Reference Point Update.
12. Solar Wind Composition (S-080).
13. Lunar Multispectral Photography (S-158).

SECONDARY DETAILED OBJECTIVES

14. Surveyor III Investigation.
15. Photographic Coverage During Lunar Landing and Lunar Surface Operations.
16. Television Coverage Through the Erectable S-band Antenna.

APOLLO 12 — Landing on the Lunar Surface

Lunar Orbit Insertion   Transfer to LEM

Lunar Descent (West Side)   Landing near Surveyor

APOLLO 12 — Lunar Surface Activities

Egress from LEM   Erection of TV camera

Contingency Sample   Bulk Sample   Erection of American flag

APOLLO 12 — Lunar Surface Activities

Astronaut carrying ALSEP 1 with 6 experiments

ALSEP 1 deployment

Documented sample collection

Walk to Surveyor

APOLLO 12 — Lunar Lift-off

Astronaut inspecting Surveyor

Return to LEM

Astronauts in LEM

Lunar lift-off

APOLLO 12 — Reentry and Recovery

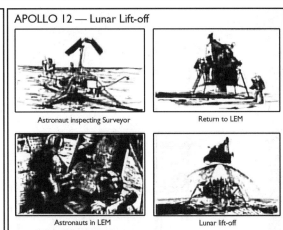

Rendezvous and Docking

Reentry and Recovery

Enter Mobile Quarantine Facility

LAUNCH COUNTDOWN AND TURNAROUND CAPABILITY, AS-507

COUNTDOWN

Countdown (CD) for launch of the AS-507 Space Vehicle (SV) for the Apollo 12 Mission will begin with a precount starting at T-98 hours during which launch vehicle (LV) and spacecraft (S/C) CD activities will be conducted independently. Official coordinated S/C and LV CD will begin at T-28 hours. Figure 2 shows the significant events beginning with the official countdown start.

SCRUB/TURNAROUND

Turnaround is the time required to recycle and count down to launch (T-0) in a subsequent launch window. The following launch window constraints apply:

* 56 hours 09 minutes are available for turnaround between the opening of the 14 November and the closing of the 16 November launch windows.

* 29 hours 13 minutes are available for turnaround between the opening of the 14 December and the closing of the 15 December launch windows.

Scrub can occur at any point in the CD when launch support facilities, SV conditions, or weather warrant. For a hold that results in a scrub prior to T-22 minutes, turnaround procedures are initiated from the point of hold. Should a hold occur from T-22 minutes (S-II start bottle chilldown) to T-16.2 seconds (S-IC forward umbilical disconnect), then a recycle to T-22 minutes, a hold, or a scrub is possible under conditions stated in the Launch Mission Rules. A hold between T-16.2 seconds and T-8.9 seconds (ignition) could result in either a recycle or a scrub depending upon the circumstances. An automatic or manual cutoff after T-8.9 seconds will result in a scrub.

Two basic cases can be identified to implement the required turnaround activities in preparation for a subsequent launch attempt following a scrub prior to ignition command. These cases identify the turnaround activities necessary to maintain the same confidence for subsequent launch attempts as for the original attempt. The scrub/turnaround time for each case is the minimum time required to effect recycle and CD of the SV to T-0 (liftoff) after a scrub. They do not account for serial times which may be required for repair or retest of any systems which may have caused the scrub, nor do they include built-in holds for launch window

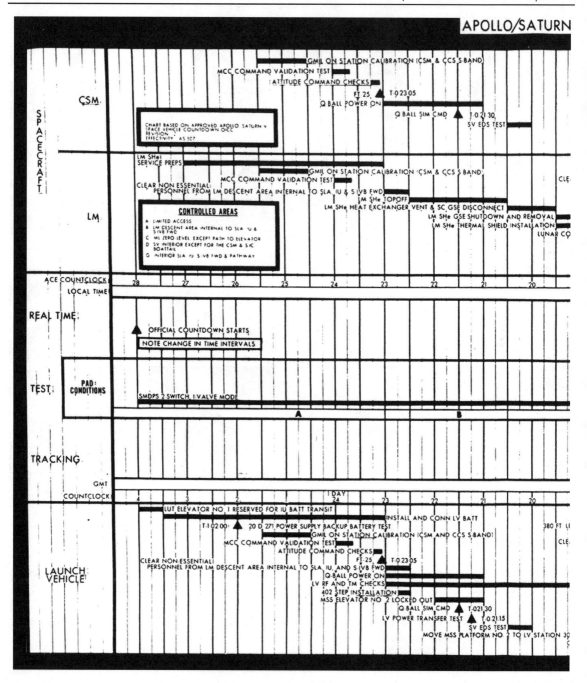

synchronization. The basic difference in the two cases is the requirement to reservice the spacecraft cryogenics, which necessitates detailed safety precautions and the reuse of the Mobile Service Structure.

48-Hour Scrub/Turnaround

A 48-hour scrub/turnaround capability exists from any point in the launch CD up to T-8.9 seconds. This turnaround capability provides for reservicing all SV cryogenics and resumption of the CD at T-9 hours.

24-Hour Scrub/Turnaround

A 24-hour turnaround capability exists as late in the CD as T-8.9 seconds. This capability depends upon having sufficient S/C consumables margins above redline quantities stated in the Launch Mission Rules (or negotiated changes to these redline quantities) for the period remaining to the next launch window. The

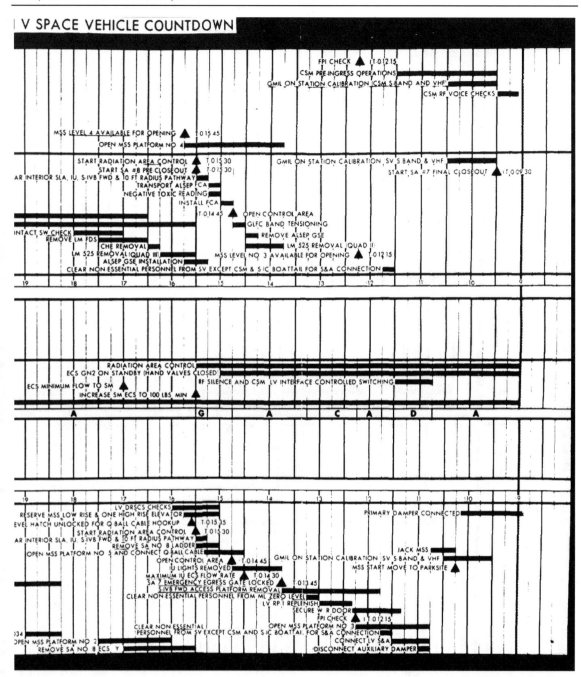

Fig. 2

CD would be resumed at T-9 hours.

FLIGHT MISSION DESCRIPTION

LANDING SITES

Apollo Lunar Landing Site 7 is the prime site for the Apollo 12 Mission. This site is located entirely within relatively old (Imbrian) mare material and also shares the characteristic distribution of large subdued 200-600 meter (m) diameter craters as well as the characteristic lower density of 50-200 m diameter craters. This site includes the crater in which Surveyor III landed in April 1967. One of the primary scientific objectives of landing at this site is to sample a second mare for comparison with Apollo 11 and Surveyor data in order to learn the variability in composition and age of the "Imbrium" mare unit.

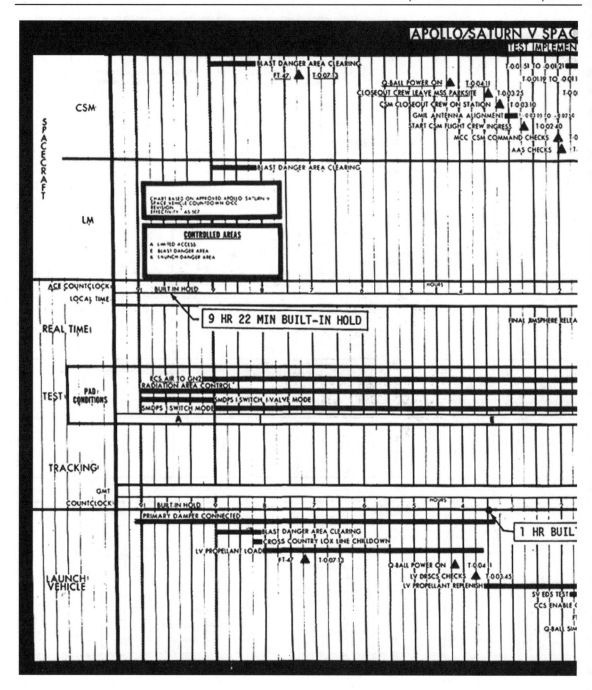

Apollo Lunar Landing Site 5 is the recycle site for this mission and is located within relatively young (Eratosthenion) mare material. In contrast to Tranquillity Base and Landing Site 7, the area of this site displays a large number of intermediate size craters 50-200 m in diameter and a small number of larger subdued craters 200-600 m in diameter. The site is surrounded by well-developed crater clusters of the Kepler system. Small, weakly developed crater clusters and lineaments radial to Kepler occur within the site. Thus some material derived from depth at Kepler may be presented in the surficial material, and fine-scale textural details related to the Kepler rays may also be present. There are more resolvable blocks (greater than 2 m) around craters than in the three sites to the east (Landing Sites 1, 2, and 3) suggesting that the surficial material is generally coarser grained.

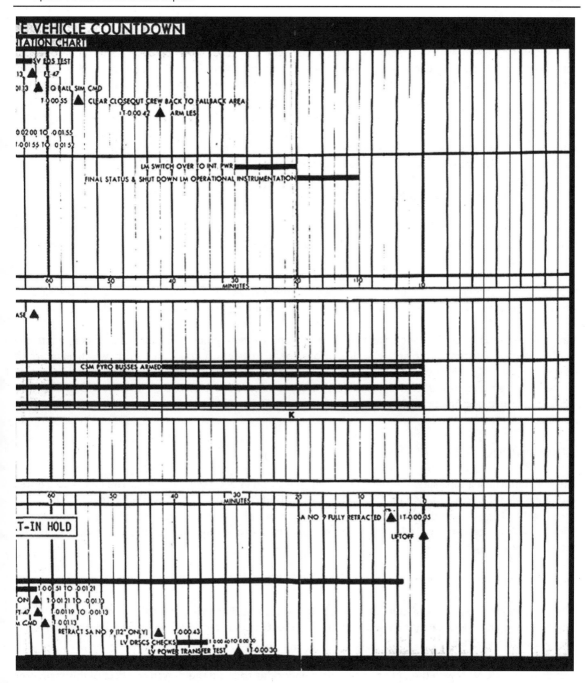

LAUNCH WINDOWS

The launch windows for both Site 7 and Site 5 are shown in Table I.

			NOV (EST)			DEC (EST)		
SITE	**LONG.**	**LAT.**	**DATE**	**OPEN-CLOSE**	**SEA***	**DATE**	**OPEN-CLOSE**	**SEA**
7	23°24'W.	2°59'S.	14	1122-1428	5.1°	14	1334-1658	10°
5	41°54'W.	1°41'N.	16	1409-1727	10.7°	15	1513-1847	5.3°

TABLE 1
APOLLO 12 LANDING SITES/LAUNCH WINDOWS

* SUN ELEVATION ANGLE

HYBRID TRAJECTORY

The Apollo 12 Mission will use a hybrid trajectory that retains most of the safety features of the free-return trajectory, but without the performance limitations. The spacecraft will be injected into a highly eccentric elliptical orbit (perilune altitude of approximately 1850 nautical miles (NM), which has the free-return characteristic, i.e., the spacecraft can return to the entry corridor without any further maneuvers. The spacecraft will not depart from the free-return ellipse until after the Lunar Module (LM) has been extracted from the launch vehicle and can provide a propulsion system backup to the Service Propulsion System (SPS). After approximately 28 hours from translunar injection, a midcourse maneuver will be performed by the SPS to place the spacecraft on a lunar approach trajectory (non-free-return) having a lower perilune altitude.

The use of a hybrid trajectory will permit:

* Daylight launch/Pacific injection. This would allow the crew to acquire the horizon as a backup attitude reference during high altitude abort, would provide launch abort recovery visibility, and would improve launch photographic coverage.

* Desired lunar landing site sun elevation. The hybrid profile facilitates adjustment of translunar transit time which can be used to control sun angles on the landing site during lunar orbit and on landing.

* Increased spacecraft performance. The launch vehicle energy requirements for translunar injection into the highly eccentric elliptical orbit are less than those for a free-return trajectory from which lunar orbit insertion would be performed. This allows for an increase in spacecraft payload/SPS propellant. The energy of the spacecraft on a hybrid lunar approach trajectory is relatively low compared to what it would be on a full free-return trajectory thus reducing the differential velocity (Delta-V) required to achieve lunar orbit insertion.

LUNAR MODULE POINT LANDING

The LM point landing capability of Apollo 12 is being enhanced in two significant areas. The first is concerned with improving the ground targeting of the Primary Guidance Navigation and Control System (PGNCS), i.e., updating the LM guidance computer with the LM's current position and velocity, and the landing site position. The second is concerned with reducing the in-orbit perturbations during the last three orbits before descent orbit insertion.

Significant improvements in ground targeting of the PGNCS include:

* Adding one more term to the computer program coverage of the lunar potential model in the Real-Time Computer Complex. This permits a significant improvement in LM orbit determination and descent targeting during a single LM orbit.

* Updating the PGNCS with the LM's position after undocking to avoid the degrading effect of this maneuver on the LM state vector.

* Updating the LM downtrack position relative to the landing site during powered descent.

Steps taken to reduce in-orbit perturbations include:

* Water and waste dumps will be avoided 8-10 hours before landing.

* LM Reaction Control System (RCS) checkout will be done with rotational maneuvers and with cold fire instead of "nulled" translational maneuvers.

* Command/Service Module (CSM) will perform undocking maneuver.

* LM undocking will be done radially to avoid downrange Delta-V.

* Soft undocking will be performed.

* Landing gear inspection will be deleted if indications are nominal.

* CSM rather than the LM will be active in station keeping.

* CSM will perform separation maneuver.

Figure 3 shows the Apollo 12 landing site. Targeting will be to Surveyor III and manual control will be used to fly to the actual landing area.

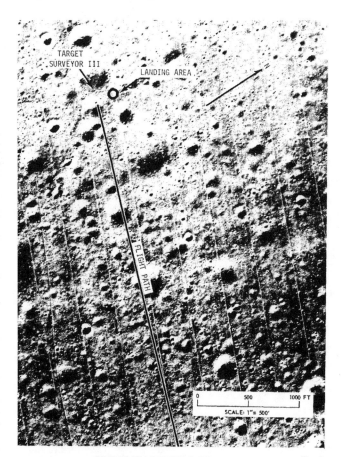

APOLLO 12 LANDING SITE   Fig. 3

FLIGHT PROFILE

Launch Through Earth Parking Orbit

The AS-507 Space Vehicle for the Apollo 12 Mission is planned to be launched at 11:22 EST on 14 November 1969 from Launch Complex 39A at the Kennedy Space Center, Florida, on a launch azimuth of 72°. The Saturn V Launch Vehicle will insert the S-IVB/Instrument Unit (IU)/LM/CSM into a 103-NM, circular orbit. The S-IVB and spacecraft checkout will be accomplished during the orbital coast phase. Figure 4 and Tables 2 through 4 summarize the flight profile events and space vehicle weight.

Translunar Injection

Approximately 2.8 hours after liftoff, the launch vehicle S-IVB stage will be reignited during the second parking orbit to perform the translunar injection (TLI) maneuver, placing the spacecraft on a free-return trajectory having a perilune of approximately 1850 NM.

Translunar Coast

The CSM will separate from the S-IVB/IU/LM approximately 3.2 hours Ground Elapsed Time (GET), transpose, dock, and initiate ejection of the LM. During these maneuvers, the LM and SIVB/IU will be photographed to provide engineering data.

An S-IVB evasive maneuver will be initiated by ground command approximately 1.6 hours after TLI. This maneuver will be performed by the Auxiliary Propulsion System (APS) of the S-IVB to impart a Delta-V of approximately 10 feet per second (fps) and prevent recontact with the spacecraft. Shortly thereafter, an S-IVB "slingshot" maneuver will be performed to place the SIVB/IU onto a trajectory passing the moon's trailing edge and into solar orbit. This maneuver will be performed by a combination of continuous hydrogen venting, liquid oxygen (LOX) dumping, and an APS ullage maneuver. The total Delta-V imparted to the S-IVB/IU by the "slingshot" maneuver will be approximately 115 fps.

The spacecraft will be placed on a hybrid trajectory by performing an SPS maneuver at the time scheduled for the second midcourse correction (MCC) approximately 31 hours from liftoff. The CSM/LM combination will be targeted for a pericynthion altitude of 60 NM and, as a result of the SPS maneuver, will be placed on a non-free-return trajectory. The spacecraft will remain within a LM Descent Propulsion System (DPS) return capability.

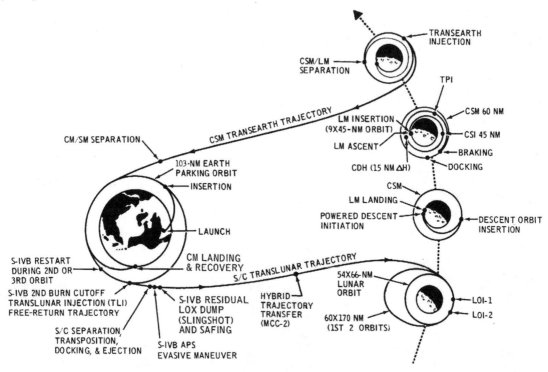

Fig. 4

APOLLO 12 FLIGHT PROFILE

TABLE 2

APOLLO 12 MISSION SUMMARY

EVENT	GET DAY:HR:MIN	DATE/EST HR:MIN	BURN DURATION (APPROX. SEC.)	REMARKS
LAUNCH	0:00:00	14/11:22		WINDOW CLOSES 1428 EST
TRANSLUNAR INJECTION (TLI)	0:02:47	14/14:09	345 (S-IVB)	PACIFIC OCEAN
MIDCOURSE CORRECTION (MCC-2)	1:06:53	15/18:15	10 (SPS)	HYBRID TRANSFER
LUNAR ORBIT INSERTION (LOI-1)	3:11:25	17/22:47	355 (SPS)	ORBIT: 59 X 169 MILES
LUNAR ORBIT INSERTION (LOI-2)	3:15:44	18/03:06	18 (SPS)	ORBIT: 53 X 65-MILES
UNDOCK	4:11:54	18/23:16	16 (CSM-RCS)	
DESCENT ORBIT INSERTION (DOI)	4:13:23	19/00:45	28 (DPS)	LM ORBIT: 59X8 MILES
POWERED DESCENT INITIATION (PDI)	4:14:20	19/01:42	679 (DPS)	
LANDING	4:14:31	19/01:53		
BEGIN EXTRAVEHICULAR ACTIVITY (EVA-1)	4:18:33	19/05:55		3 HOURS 30 MINUTES
BEGIN EVA-2	5:13:07	20/00:29		3 HOURS 30 MINUTES
LM LIFTOFF	5:22:01	20/09:23	430 (APS)	LM ORBIT 8 X 45 MILES
DOCKING	6:01:40	20/13:02		
LUNAR ORBIT PLANE CHANGE	6:15:02	21/02:24	18 (SPS)	
TRANSEARTH INJECTION (TEI)	7:04:21	21/15:43	129 (SPS)	
LANDING	10:04:35	24/15:57		LATITUDE = 16°S LONGITUDE = 165°W LOCAL TIME 09:57 (SUN RISE + 5 HR.)

MISSION DURATION: 244 HR. 35 MIN.

TABLE 3

APOLLO 12 TV SCHEDULE

DAY	DATE	EST	GET	COVERAGE
FRIDAY	NOV. 14	14:42	03:25	TRANSPOSITION / DOCKING
SATURDAY	NOV. 15	17:47	30:25	HYBRID TRAJ. / SPACECRAFT INTERIOR
MONDAY	NOV. 17	02:52	153:30	EARTH, IVT, S/C INTERIOR
		20:52	81:30	PRE LOI-1, LUNAR SURFACE
		23:22	84:00	LUNAR SURFACE
TUESDAY	NOV. 18	23:12	107:50	UNDOCKING / FORMATION FLYING
WEDNESDAY	NOV. 19	06:02	114:40	LUNAR SURFACE EVA
THURSDAY	NOV. 20	00:42	133:20	EVA - 2, EQUIPMENT JETTISON
		12:37	145:15	DOCKING
FRIDAY	NOV. 21	16:17	172:55	POST - TEI / LUNAR SURFACE
SUNDAY	NOV. 23	18:37	223:15	MOON - EARTH - S/C INTERIOR

TABLE 4

APOLLO 12 WEIGHT SUMMARY (weight in Pounds)

STAGE/MODULE	INERT WEIGHT	TOTAL EXPENDABLES	TOTAL WEIGHT	FINAL SEPARATION WEIGHT
S-IC Stage	287,850	4,742,865	5,030,715	363,465
S-IC/S-II Interstage	11,465	—	11,465	—
S-II Stage	80,220	980,200	1,060,420	94,440
S-II/S-IVB Interstage	8,035	—	8,035	—
S-IVB Stage	25,050	235,020	262,070	28,440
Instrument Unit	4,275	—	4,275	—
	Launch Vehicle at Ignition		6,374,980	
Spacecraft-LM Adapter	4,060	—	4,060	—
Lunar Module	9,635	23,690	33,325	*33,740
Service Module	10,510	40,595	51,105	11,840
Command Module	12,365	—	12,365	11,145 (Landing)
Launch Escape System	8,945	—	8,945	—
	Spacecraft At Ignition		109,800	
Space Vehicle at ignition			6,484,780	
S-IC Thrust Buildup			(-)85,320	
Space Vehicle at Liftoff			6,399,300	
Space Vehicle at Orbit Insertion			300,269	

*CSM/LM Separation.

The earth will be photographed several times each day during this coast phase for oceanographic, global weather, and documentation purposes as the spacecraft attitude and crew time permit. The moon will also be photographed. MCC's will be made as required, using the Manned Space Flight Network (MSFN) for navigation.

Lunar Orbit Insertion

The SPS will insert the spacecraft into an initial lunar orbit (approximately 60 x 170 NM) 83.4 hours from liftoff (Figure 4). Following insertion and systems checks, 0 second SPS retrograde burn will be made to place the spacecraft in an elliptical orbit 54 x 66 NM. This orbit is planned to become circular at 60 NM by the time of LM rendezvous.

Because lunar orbit insertion (LOI) always occurs behind the moon, the crew will be required to evaluate the progress of the maneuver without ground support. Although two LOI burns are required to produce a near circular orbit, the monitoring requirements primarily impact the first burn (LOI-I), because the second burn (LOI-2) lasts for only approximately 18 seconds. The horizon and several stars should be visible from the Commander's (CDR's) rendezvous window and may be used as a backup to the optics for the orientation check prior to SPS ignition.

Lunar Orbit Coast

After LOI-2 the Lunar Module Pilot (LMP) and the CDR will enter the LM to perform housekeeping and the initial LM activation. Subsequently, a rest and eat period of approximately 8.5 hours will be provided for the three astronauts prior to LM activation and checkout.

The CSM will separate radially upward from the LM at approximately 20.7 hours from LOI-2 using the soft undocking technique. The docking probe capture latches will be used to minimize separation Delta-V perturbations. After undocking, the CSM will maintain a distance of 40 feet from the LM. The LM will not perform any inspection maneuvers (e.g., landing gear inspection), unless there is a real-time indication that the landing gear did not deploy properly.

Lunar Module Descent

The DPS will be used to perform the descent orbit insertion (DOI) maneuver approximately 1.5 hours after CSM/LM separation. This maneuver places the LM in a 60-NM by 50,000-foot orbit as shown in Figure 5.

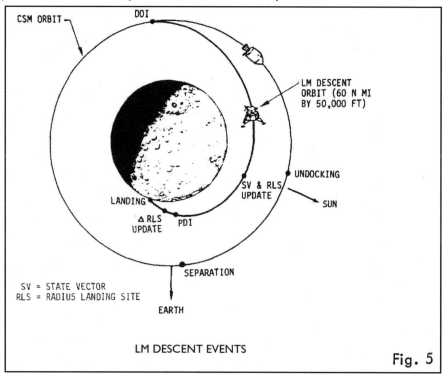

LM DESCENT EVENTS

Fig. 5

Powered descent initiation (PDI) will occur near the pericynthion of the descent orbit (Figure 6). The vertical descent portion of the landing phase will start at an altitude of approximately 100 feet for an automatic approach. Present plans provide for manual takeover by the crew at an altitude of 500 feet.

During descent the lunar surface will be photographed to record LM movement, surface disturbances, and to aid in determining the landed LM location.

Lunar Surface Operations

Post landing

Immediately upon landing, the LM crew will execute the lunar contact checklist and reach a stay/no-stay decision. After reaching a decision to stay, the Inertial Measurement Unit will be aligned, the Abort Guidance System gyro calibrated and aligned, and the lunar surface photographed through the LM window. Following a crew eat period all loose items not required for extravehicular activity (EVA) will be stowed.

EVA I

The activity timeline for EVA I is shown in Figure 7. Both crew members will don helmets, gloves, Portable Life Support Systems (PLSS), and Oxygen Purge Systems (OPS) and the cabin will be depressurized from 3.5 pounds per square inch (psi).

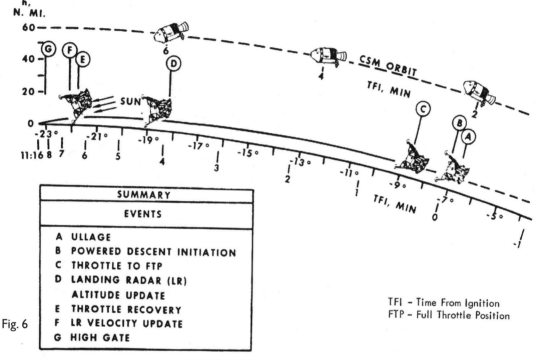

Fig. 6

SUMMARY EVENTS

A ULLAGE
B POWERED DESCENT INITIATION
C THROTTLE TO FTP
D LANDING RADAR (LR) ALTITUDE UPDATE
E THROTTLE RECOVERY
F LR VELOCITY UPDATE
G HIGH GATE

TFI – Time From Ignition
FTP – Full Throttle Position

LM POWERED DESCENT

The CDR will move through the hatch, deploy the Lunar Equipment Conveyor (LEC), and move to the ladder where he will deploy the Modularized Equipment Stowage Assembly (MESA), Figure 8, which initiates television coverage from the MESA. He will then descend the ladder to the lunar surface. The LMP will monitor and photograph the CDR using a 70mm and a sequence camera (16mm Data Acquisition Camera).

Environmental Familiarization/Contingency Sample Collection - After stepping to the surface and checking his mobility, stability, and the Extravehicular Mobility Unit (EMU), the CDR will collect a contingency sample. This would make it possible to assess the differences in the lunar surface material between the Apollo 11 and 12 landing sites in the event the EVA were terminated at this point. The sample will be collected by quickly scooping up a loose sample of the lunar material (approximately 2 pounds), sealing it in a Contingency Sample Container, and transferring the sample in the Equipment Transfer Bag (ETB) along with the lithium hydroxide (LiOH) canisters and PLSS batteries into the LM using the LEC. The LMP will then transfer the 70mm cameras to the surface in the ETB.

Contingency Photography - The CDR will photograph the contingency sample area, deploy and photograph the color chart in the sunlight, and photograph the descent of the LMP to the surface.

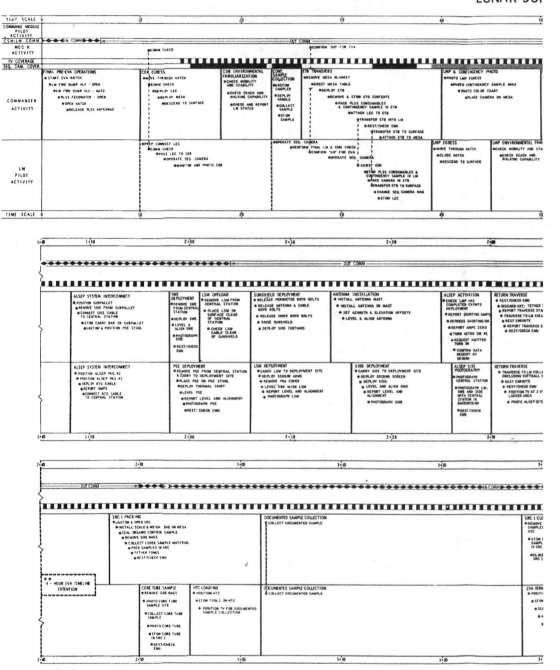

S-band Antenna Deployment - The S-band antenna will be removed from the LM and carried to the site where the CDR will erect it as shown in Figure 9, connect the antenna cable to the LM, and perform the required alignment.

Flag Deployment - The CDR will then unstow the American flag and carry it to the deployment site and implant it in the lunar surface.

Lunar TV Camera Deployment - While the CDR deploys the S-band antenna, the LMP will unstow the camera and deploy it on the tripod approximately 20 feet from the LM in the 10 o'clock position (Figure 10). The LMP will then obtain TV panorama and special interest views after which he will point the camera at the S-band antenna/flag deployment/MESA area.

RFACE ACTIVITY TIMELINE -EVA I

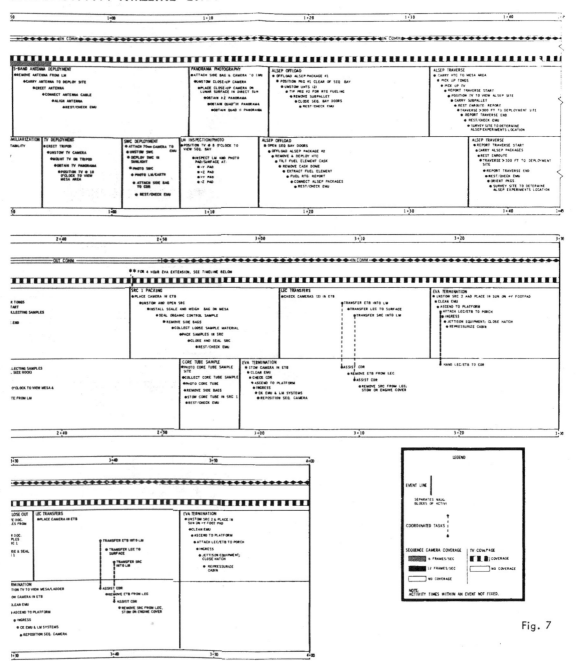

Fig. 7

Solar Wind Composition Deployment - The LMP will next unstow and deploy the Solar Wind Composition (SWC) experiment which uses a 4-square foot aluminum foil area for entrapment of solar wind particles. It will be carried to the deployment site where the foil will be unfurled and the staff implanted in the lunar surface. As in the Apollo 11 Mission, the SWC detector will be brought back to earth by the astronauts. However, on Apollo 12 the detector will be exposed to the solar wind flux for approximately 17 hours instead of 2 hours and will be placed a sufficient distance away from the LM to protect it from lunar dust kicked up by astronaut activity.

LM Inspection - After repositioning the TV to view the Scientific Equipment Bay door area, the LMP will inspect and photograph the LM footpads; and quadrants (QUAD'S) 1, II, III, and IV with his 70mm camera.

DEPLOYED MESA   Fig. 8   DEPLOYED S-BAND ANTENNA   Fig. 9

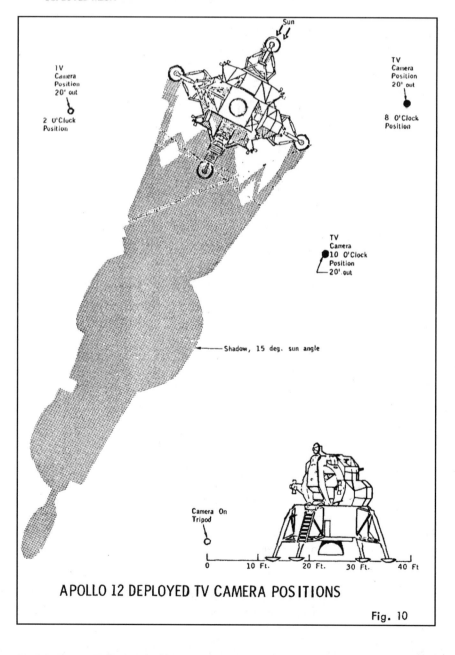

APOLLO 12 DEPLOYED TV CAMERA POSITIONS

Fig. 10

Concurrently the CDR will obtain panorama and close-up photographs.

ALSEP Deployment - Both crew members will off load, deploy, and activate the Apollo Lunar Surface Experiments Package (ALSEP) which will obtain scientific data consisting of lunar physical and environmental characteristics and transmit the data to earth for determination of (1) the magnetic fields at the moon, (2) the lunar atmosphere and ionosphere and the lunar seismic activity, and (3) the properties of the solar wind plasma as it exists at the lunar surface. The ALSEP is stowed and off loaded in two sub packages. The fuel cask (part of the electrical power subsystem) is attached to the LM.

After off loading the ALSEP packages, the Radioisotope Thermoelectric Generator (RTG), which provides the ALSEP electrical power, will be fueled (Figure 11), the ALSEP sub packages will be attached to a one-man carry bar for traverse in a "barbell" mode, as shown on the cover, and the TV will be positioned to view the ALSEP site.

The LMP will then carry the ALSEP sub packages in the "barbell" mode to the deployment site approximately 300 feet from the LM while the CDR carries a subpallet of ALSEP, Upon arriving at the deployment site they will survey the site and determine the desired location for the experiments. The following individual experiment packages will then be separated, assembled, connected to the ALSEP cabling, and deployed to their respective sites (Figure 12).

Passive Seismic Experiment (Figure 13) - This experiment is designed to monitor seismic activity and affords the opportunity to detect meteoroid impacts and free oscillations of the moon. It may also detect surface tidal deformations resulting in part from periodic variations in the strength and direction of external gravitational fields acting upon the moon.

Solar Wind Spectrometer Experiment (Figure 14) - This experiment will measure energies, densities, incidence angles, and temporal variations of the electron and proton components of the solar wind on the surface of the moon.

RADIOISOTOPE THERMOELECTRIC
GENERATOR FUELING

Fig. 11

Lunar Surface Magnetometer Experiment (Figure 15) - This experiment will measure the magnitude and temporal variations of the lunar surface equatorial field vector.

Suprathermal Ion Detector (Lunar Ionosphere Detector) Experiment (Figure 16) This experiment will measure the flux, number, density, velocity, and energy per unit charge of positive ions in the vicinity of the lunar surface.*

Cold Cathode Ion Gauge (Lunar Atmosphere Detector) Experiment - This experiment will determine the density of any lunar ambient atmosphere including variations either of a random character or associated with lunar local time or solar activity. In addition, the rate of loss of contaminants left in the landing area by the astronauts and the Lunar Module will be measured.*

* On ALSEP 1, the suprathermal ion detector and cold cathode ion gauge will be integrated together in one experiment system.

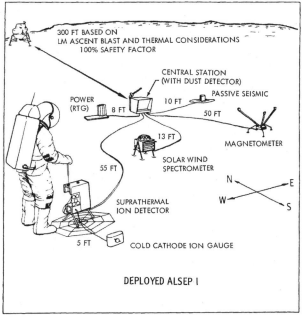

Fig. 12

Lunar Dust Detector Experiment - This experiment will obtain data for the assessment of dust accretion on ALSEP to provide a measure of the degradation of thermal surfaces.

Following the deployment of experiments, the ALSEP will be activated, data receipt by MSFN confirmed, and the ALSEP site and deployed experiments photographed. The ALSEP site will also be photographed from the LM area.

Selected Sample Collection - During the return traverse to the LM, both crewmen will collect a selected sample of geologically interesting material, including rock samples and fine grained fragmental material, which will be carried in a side bag on each crewman. Approximately three-fourths of the quantity will be rock samples with the remaining one-fourth fine-grained material. The samples and the immediate sample site will be photographed.

DEPLOYED PASSIVE SEISMIC EXPERIMENT

Fig. 13

DEPLOYED SOLAR WIND
SPECTROMETER EXPERIMENT

Fig. 14

DEPLOYED LUNAR SURFACE
MAGNETOMETER EXPERIMENT

Fig. 15

DEPLOYED SUPRATHERMAL ION DETECTOR/
COLD CATHODE ION GAUGE EXPERIMENT

Fig. 16

The LMP will carry the TV back to the LM area and position it to view the MESA and surrounding area from the 2 o'clock position shown in Figure 10. The LMP will then assemble the core tube and handle, and collect a core sample. After collecting the core sample, the sample will be capped and stowed in Sample Return Container (SRC) 1.

Upon return to the LM, the CDR will unstow the selected SRC, attach the scale to the MESA, finish filling the CDR and LMP side bags with loose material, seal the organic control sample, pack the samples, and seal the SRC.

After helping the CDR with the selected sample collection, the LMP will clean his EMU, ingress the LM, check LM systems, switch to the erectable S-band antenna, and make a communications check. The CDR will attach the LEC to the SRC 1 and transfer it into the LM with the assistance of the LMP.

Post-EVA 1 Operations

After configuring the LM systems for post-EVA 1 operations, the PLSS's will be recharged. This includes filling the oxygen system to a minimum pressure of 875 psi, filling the water reservoir, and replacing the battery and LiOH canister. The PLSS's and OPS's will be doffed and stowed, followed by an eat period, a 9-hour rest period, and another eat period.

EVA 2

After pre-EVA configuring of the EMU's and LM systems, the cabin will be depressurized from 3.5 psi and the CDR will descend to the surface for EVA 2 (Figure 17). Upon transferring the 70mm Lunar Surface Cameras to the surface using the ETB and LEC (Figure 18), and turning on the 16mm Data Acquisition Camera (Sequence Camera) in the LM, the LMP will descend to the surface.

Lunar Field Geology Experiment - Both crewmen will participate in the conduct of the Lunar Field Geology Experiment, which is to provide data for use in the interpretation of the geologic history of the moon. A team of earth-based geologists will be available to advise the astronauts in real-time.

Geology traverse preparations will include stowing several contrast charts, a hammer, an extension handle, a small and a large scoop, core tubes and caps, sample bag dispenser, and a gnomon on the Hand Tool Carrier (HTC) (Figure 19); attaching side bags; stowing the cutting tool in the CDR's Surveyor parts bag; attaching a 70mm Lunar Surface Camera to each EMU; tethering tongs to the CDR's EMU; deploying contrast charts; and repositioning the TV for geology traverse.

The geology traverse for this experiment will consist of documented sample collection, core tube sample collection, trench site sample collection, and gas analysis sample collection. A typical documented sample collection will include photographing the sample and its site and describing and stowing the sample in a sample bag. A typical core tube sample collection will include photographing the sample site cross sun, driving the core tube into the surface and photographing the core tube down sun, and pulling and capping the core tube. Trench site sample collection will include digging a trench along the sunline, filling the special environmental sample container with surface material and sealing it, photographing the trench both down and up sun, collecting a core sample from the trench, and stowing the samples in the HTC. Gas analysis sample collection will include photographing the sample both cross and down sun, collecting, the sample using tongs, and placing it in and sealing the gas analysis sample container which will be stowed in the HTC.

Surveyor site and vehicle investigation will precede the geology return traverse.

Surveyor Site Activity - As a secondary objective, it is planned that the CDR and LMP will walk to the Surveyor III site for an investigation of the site and the Surveyor vehicle (Figure 20). The CDR and LMP will descend into the crater containing the Surveyor III and collect samples of lunar material including lunar bedrock, layered rock, and rounded rocks in ray patterns. The LMP will obtain photographs of lunar material in the vicinity of and deposited on the Surveyor III spacecraft as well as several photographs of the Surveyor

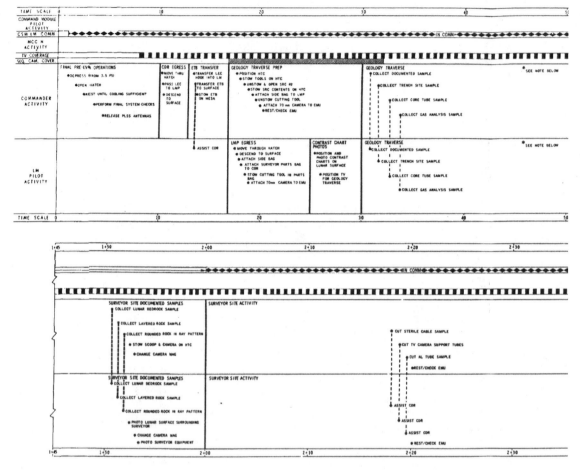

Fig. 17   LUNAR SURFACE ACTIVITY TIMELI

spacecraft equipment (Figure 21). The CDR will read the LMP's checklist during the LMP's photography and then cut the TV camera, a piece of the TV camera electrical cable which will be dropped untouched into the special environmental sample container, and a polished aluminum tube from the Surveyor using the cutting tool. The LMP will assist the CDR in the cutting task and will stow the equipment in the Surveyor parts bag on the CDR's PLSS. In addition, if feasible and safe, the CDR and LMP will collect pieces of glass from the Surveyor III spacecraft mirrors and report on the extent of debonding.

Post-Geology Activity - After completion of the geology return traverse, the TV will be repositioned to view the MESA and the ladder; the SWC will be camera will be retrieved and stowed in the SRC; the 70mm lunar surface cameras will be stowed in the ETB; the side bag samples, the core tubes, special environmental sample container, gas analysis sample container, and documented samples will be transferred in the SRC (Figure 22); and the LMP will obtain surface close-up photographs with the Apollo Lunar Surface Close-up Camera

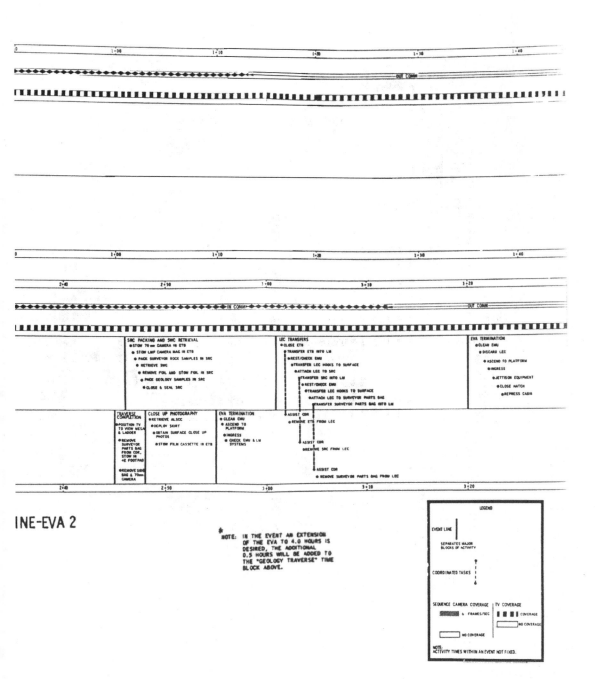

INE-EVA 2

NOTE: IN THE EVENT AN EXTENSION OF THE EVA TO 4.0 HOURS IS DESIRED, THE ADDITIONAL 0.5 HOURS WILL BE ADDED TO THE "GEOLOGY TRAVERSE" TIME BLOCK ABOVE.

(Figure 23). After both EMU's have been checked and cleaned, the LMP will ingress the LM and assist the CDR in transferring the SRC, ETB, and the Surveyor parts bag into the LM.

EVA-2 Termination - After completing equipment transfer to the LM, the CDR will clean his EMU, ascend the ladder and ingress into the LM. Expendable equipment will be jettisoned and the cabin repressurized terminating the second EVA.

CSM Lunar Orbit Operations

Lunar Multispectral Photography

During the period of lunar surface operations, the Command Module Pilot (CMP) will obtain simultaneous

EQUIPMENT TRANSFER BAG/
LUNAR EQUIPMENT CONVEYOR

Fig. 18

HAND TOOL CARRIER

Fig. 19

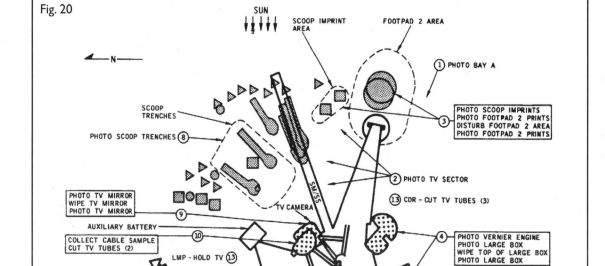

Fig. 20

SURVEYOR III ACTIVITIES

multispectral photographs of the lunar surface at three widely separated wavelengths. This photography will provide data on lunar surface color variations (at an order of magnitude higher resolution than obtainable from earth) which will be useful in geologic mapping. For example, the sharpness of the color boundaries will give a good indication of the compositional differences. In addition, it will provide data for correlation with the spectral reflectance properties of the returned lunar samples from Apollo I I and thus will allow possible extrapolation of compositional information on other areas of the moon on which no landings will occur. Finally, it will define areas of interest for future correlation with the returned samples.

SAMPLE RETURN CONTAINER

Fig. 22

SURVEYOR III

Fig. 21

LUNAR SURFACE CLOSE-UP CAMERA

Fig. 23

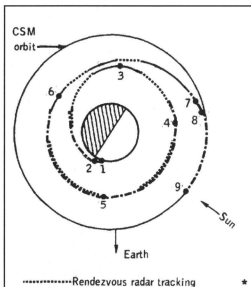

EVENT
1. LIFTOFF
2. INSERTION
3. CSI
4. PLANE CHANGE
5. CDH
6. TPI
7. BEGIN BRAKING*
8. STATION KEEPING
9. DOCKING

CSM orbit

Sun

Earth

............Rendezvous radar tracking
—·—·—·MSFN tracking

* A TOTAL OF 4 BRAKING MANEUVERS ARE PLANNED.

LM ASCENT THROUGH DOCKING

Fig. 24

Lunar Module Ascent

The LM ascent (Figure 24) will begin after a lunar stay of approximately 31.5 hours. The Ascent Propulsion System (APS) powered ascent is divided into two phases. The first phase is a vertical rise which is required to achieve terrain clearance, and the second phase is orbit insertion. After orbit insertion the LM will execute the coelliptic rendezvous sequence which nominally consists of four major maneuvers: concentric sequence initiation (CSI), constant delta height (CDH), terminal phase initiation (TPI), and terminal phase finalization (TPF). A nominally zero plane change maneuver will be scheduled between CSI and CDH, and two nominally zero midcourse correction maneuvers will be scheduled between TPI and TPF; the TPF maneuver is actually divided into several braking maneuvers. All maneuvers after orbit insertion will be performed with the LM RCS. Once docked to the CSM, the two crewmen will transfer to the CSM with equipment, lunar samples, Surveyor III parts, and exposed film. Decontamination operations will be performed, jettisonable items will be placed in the Interim Stowage Assembly and transferred to the LM, and the LM will be configured for deorbit and lunar impact.

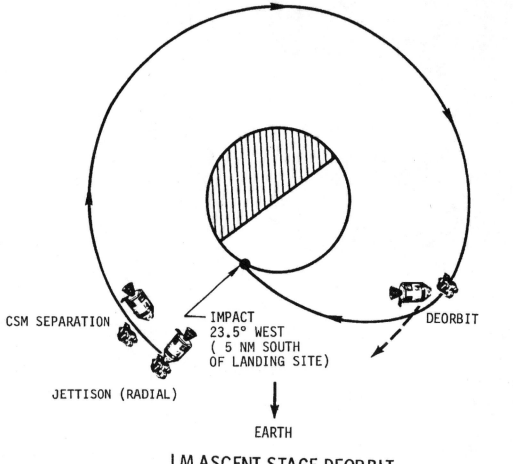

CSM SEPARATION

IMPACT
23.5° WEST
(5 NM SOUTH
OF LANDING SITE)

DEORBIT

JETTISON (RADIAL)

EARTH

LM ASCENT STAGE DEORBIT

Fig. 25

LM Ascent Stage Deorbit

The ascent stage will be deorbited for lunar surface impact near the newly deployed ALSEP, rather than sent into solar orbit, to provide a known perturbation for the seismic experiment. (Figure 25). The CSM in a heads-up attitude will be separated radially from the ascent stage with a Service Module (SM) RCS retrograde burn approximately 2 hours after docking to the CSM. Following the LM jettison maneuver, the CSM will perform a pitch down maneuver. The LM deorbit maneuver will be a retrograde APS burn initiated by ground control and the LM will be targeted to impact the lunar surface approximately 5 NM south of the Apollo 12 landing site. The ascent stage jettison, ignition, and impacted lunar surface area will be photographed from the CSM.

CSM Orbit Operations

Photography of Candidate Exploration Sites

After ascent stage deorbit the CSM will execute an orbital plane change for approximately 11 hours of lunar reconnaissance photography. Stereoscopic and sequence photographs in high resolution will be taken of Descartes, Fra Mauro, LaLande, and other candidate sites, as feasible, prior to trans-earth injection.

Trans-earth Injection and Coast

The SPS will be used to inject the CSM onto the trans-earth trajectory. The return flight duration will be approximately 72 hours (based on a 14 November launch) and the return inclination (to the earth's equator) will not exceed 40 degrees. Midcourse corrections will be made as required, using the MSFN for navigation.

Entry and Recovery

Prior to atmospheric entry, the CSM will maneuver to a heads-up attitude, the Command Module (CM) will jettison the SM and orient to the entry attitude (heads down, full lift). The nominal range from entry interface (EI) at 400,000 feet altitude to landing will be approximately 1250 NM. Earth landing will nominally be in the Pacific Ocean at 16°S latitude and 165°W longitude (based on a 14 November launch) approximately 244.6 hours after liftoff. Immediate recovery is planned.

Quarantine

Following landing, the Apollo 12 crew will don the flight suits and face masks passed in to them through the spacecraft hatch by a recovery swimmer wearing standard scuba gear. The flight suit/oral-nasal mask combination will be used in lieu of the integral Biological Isolation Garments (BIG's) used on Apollo 11. The BIG's will be available for use in case of an unexplained crew illness. The swimmer will swab the hatch and adjacent areas with a liquid decontamination agent. The crew will then be carried by helicopter to the recovery ship where they will enter a Mobile Quarantine Facility (MQF) and all subsequent crew quarantine procedures will be the same as for the Apollo 11 Mission.

The spacecraft will be returned to port by the recovery ship where a team will deactivate pyrotechnics, and flush and drain fluid systems (except water) - This operation will be confined to the exterior of the spacecraft. The spacecraft will then be flown to the Lunar Receiving Laboratory (LRL) and placed in a special room for storage. Lunar sample release from the LRL is contingent upon spacecraft sterilization. Contingency plans call for sterilization and early release of the spacecraft if the situation so requires.

TABLE 5
COMPARISON OF MAJOR DIFFERENCES
APOLLO 11 vs. APOLLO 12

EVENT	APOLLO 11	APOLLO 12
1. LAUNCH AZIMUTH	72 - 108°	72 - 96°
2. TRAJECTORY	FREE-RETURN	HYBRID
3. EVASIVE MANEUVER	CSM	S-IVB APS
4. NAVIGATION		PROCEDURAL CHANGES
5. EVA	1:(2 HR 32 MIN)	2:(3 HR 30 MIN EACH)
6. EVA RADIUS (MAX)	250 FT	OPS PURGE CAPABILITY
7. LUNAR SURFACE STAYTIME	21.6 HR	~31.5 HR
8. LUNAR ORBIT STAYTIME	59.6 HR	~89 HR
9. EXPERIMENTS	EASEP	ALSEP
10. PHOTOGRAPHY		MULTISPECTRAL TERRAIN 500MM LENS LUNAR LANDING SITES
11. SLEEPING (LM)		HAMMOCK ARRANGEMENT
12. LUNAR SURFACE TV	BLACK & WHITE	COLOR
13. ASCENT STAGE	IN ORBIT	DEORBIT
14. TRANSEARTH FLIGHT	59.4 HR	72.2 HR
15. TOTAL MISSION TIME	195.3 HR	244.6 HR

Apollo 11/12 Mission Differences
The major differences between the Apollo 11 and 12 flight missions are summarized in Table 5.

CONTINGENCY OPERATIONS

GENERAL

If an anomaly occurs after liftoff that would prevent the space vehicle from following its nominal flight plan, an abort or an alternate mission will be initiated. Aborts will provide for an acceptable flight crew and Command Module (CM) recovery while alternate missions will attempt to maximize the accomplishment of mission objectives as well as provide for an acceptable flight crew and CM recovery. Figure 26 shows the Apollo 12 contingency options.

ABORTS

The following sections present the abort procedures and descriptions in order of the mission phase in which they could occur.

Launch

There are six launch abort modes. The first three abort modes would result in termination of the launch sequence and a CM landing in the launch abort area. The remaining three abort modes are essentially alternate launch procedures and result in insertion of the Command/Service Module (CSM) into earth orbit. All of the launch abort modes are the same as those for the Apollo 11 Mission.

Earth Parking Orbit

A return to earth abort from earth parking orbit (EPO) will be performed by separating the CSM from the remainder of the space vehicle and performing a retrograde Service Propulsion System (SPS) burn to effect entry. Should the SPS be inoperable, the Service Module Reaction Control System (SM RCS) will be used to perform the deorbit burn. After CM/SM separation and entry, the crew will fly a guided entry to a pre-selected target point, if available.

Translunar Injection

Translunar injection (TLI) will be continued to nominal cutoff, whenever possible, in order for the crew to perform malfunction analysis and determine the necessity of an abort.

Translunar Coast

If ground control and the spacecraft crew determine that an abort situation exists, differential velocity (Delta-V) targeting will be voiced to the crew or an onboard abort program will be used as required. In most cases, the Lunar Module (LM) will be jettisoned prior to the abort maneuver if a direct return is required. An SPS burn will be initiated to achieve a direct return to a landing area. However, a real-time decision capability will be exploited as necessary for a direct return or circumlunar trajectory by use of the several CSM/LM propulsion systems in a docked configuration.

For a nominal spacecraft trajectory, an abort at TLI plus 90 minutes will require approximately 5160 feet per second Delta-V to return the spacecraft to a contingency landing area.

Lunar Orbit Insertion

An early shutdown of the SPS may result from a manual shutdown due to critical SPS problems or from an inadvertent shutdown. If an inadvertent shutdown occurs early in the first lunar orbit insertion (LOI) burn, an immediate restart of the SPS should be attempted provided specified performance "limits" are not exceeded. If restart of the SPS is not required, the LM Descent Propulsion System (DPS) will be the primary abort propulsion system. The LM Ascent Propulsion System (APS) will be required to supplement the DPS in order to meet the propulsion requirements of some abort conditions when a hybrid trajectory is used.

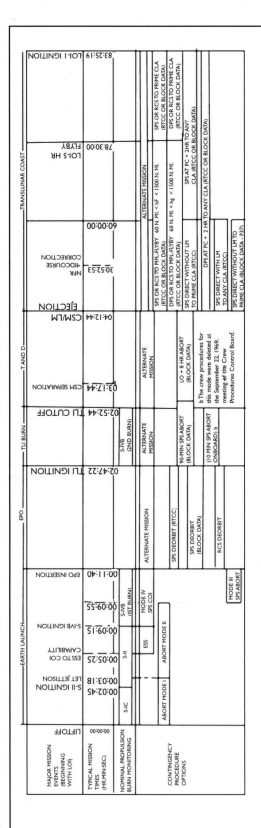

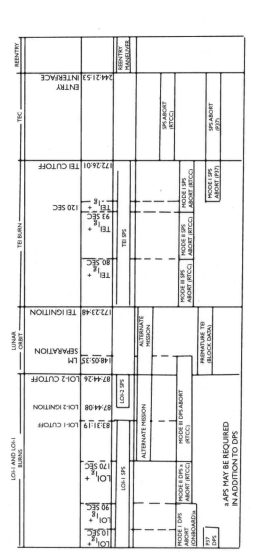

APOLLO 12 CONTINGENCY OPTIONS

Fig. 26

Mode I (LOI-1 ignition to 90 seconds): Initiate a DPS abort at 30 minutes after LOI ignition. If a satisfactory trans-earth coast is not achieved because of DPS Delta - V limitations, initiate an SPS burn 2.5 hours after LOI ignition. If the SPS is not available, the APS should be used.

Mode II (90-170 seconds after LOI-1 ignition): Initiate a DPS first burn under Real-Time Computer Complex (RTCC) control 2 hours after LOI-1 ignition. Initiate a DPS second burn after one revolution in an intermediate ellipse. Between 90 and 144 seconds after the LOI burn, the second DPS burn will be followed by an APS burn to inject the spacecraft into the desired trans-earth trajectory.

Mode III (170 seconds to end of LOI): Initiate a DPS abort (RTCC) after one revolution.

Trans-earth Injection

An SPS shutdown during trans-earth injection (TEI) may occur as the result of an inadvertent automatic shutdown. Manual shutdowns are not recommended. If an automatic shutdown occurs, an immediate restart will be initiated. If immediate reignition of the SPS is not possible, the following aborts apply if the SPS problems can be resolved.

Mode I (93 seconds to end of TEI burn): Initiate one SPS burn 2 hours after TEI ignition The preabort trajectory will be a hyperbola.

Mode II (80 to 93 seconds into TEI burn): Two SPS burns are required. Initiate the first burn 2 hours after TEI ignition. The preabort trajectory will be a hyperbola.

Mode III (TEI ignition to 80 seconds): Initiate one SPS burn after one or more revolutions. The preabort trajectory will be a stable ellipse.

ALTERNATE MISSION SUMMARY

The two general categories of alternate missions that can be performed during the Apollo 12 Mission are (1) earth orbital, and (2) lunar. Both of these categories have several variations which depend upon the nature of the anomaly causing the alternate mission and the resulting systems status of the LM and CSM. A brief description of these alternate missions is contained in the following paragraphs.

Earth Orbital Alternate Missions

Contingency: No TLI or partial TLI.

Alternate Mission: The first day in earth orbit will consist of extraction and crew entry of the LM, separation of the CSM and S-IVB maneuver, and performance of a photographic mission in the CSM/LM docked configuration.

During the second day, the LM will be deorbited for ocean impact and a CSM plane change along with a maneuver to achieve an elliptical orbit will be made. The photographic mission will continue during the third through the fifth day in orbit. If the photographic mission is complete by 100 hours Ground Elapsed Time (GET), the spacecraft will enter and land.

Lunar Orbit Alternate Missions

Contingency: Failure to eject LM from S-IVB.

Alternate Mission: Perform landmark tracking and photographic mission with the CSM with special emphasis on obtaining photographs of the bootstrap sites of the nominal mission.

The first activity day in lunar orbit will consist of LOI-1, LOI-2, landmark tracking, and high resolution and

TABLE 6

NETWORK CONFIGURATION FOR APOLLO 12 MISSION

Systems	TRACKING		USB					TLM				DATA PROCESSING				COMM				OTHER			REMARKS
	C-band (High Speed)	C-band (Low Speed)	USB	TV to MCC	Voice	TLM	Command	VHF Links	FM Remoting	Mag Tape Recording	Decoms	642B TLM	642B CMD	CDP	1218	High Speed Data	TTY	Voice (SCAMA)	Video VHF A/G	Range Safety	Riometer	SPAN	
ACN			X		X	X	X	X	X	X	X	X	X			X	X	X	X				
ANT																				X			
ARIA(4)			X		X	X		X		X							X	X	X				Note 1
ASC																				X			
AOCC																	X		X				
BDA	X	X	X		X	X	X	X	X	X	X	X	X			X	X	X	X	X			
CNV																				X			
CRO	X	X	X		X	X	X	X	X	X	X	X	X			X	X	X	X		X	X	
CYI			X		X	X	X	X	X	X	X	X	X			X	X	X	X		X	X	
GBI																				X			
GDS			X	X	X	X		X	X	X	X	X	X			X	X	X	X				
GDS-X			X		X	X	X									X							Note 2
GTK																				X			
GWM			X		X	X	X	X	X	X	X	X	X			X	X	X	X	X			
GYM			X		X	X	X	X	X	X	X	X	X			X	X	X	X				
HAW			X		X	X	X	X	X	X	X	X	X			X	X	X	X	X			
HSK			X	X	X	X		X	X	X	X	X	X			X	X	X					
HSK-X			X		X	X	X									X							Note 2
MAD			X	X	X	X		X	X	X	X	X	X			X	X	X	X				
MAD-X			X		X	X										X							Note 2
MARS				X	X	X		X															Note 3
MIL			X		X	X	X	X	X	X	X	X	X			X	X	X	X				
MLA	X																			X			
PAT																				X			
TEX			X		X	X	X	X	X	X	X	X	X			X	X	X	X				
VAN	X	X	X		X	X	X	X	X	X	X	X	X		X	X	X	X	X	X			
PARKES				X	X	X		X															NOTE 3

Note

1. TLI and reentry
2. Post TLI coverage
3. Lunar surface operations only

vertical stereo photography followed by a 6-hour steep cycle.

The second activity day will consist of two plane changes with landmark tracking, vertical stereo photography, and high resolution photography of selected science sites followed by a 10-hour sleep cycle.

The third activity day will consist of one plane change, landmark tracking, vertical stereo photography, high resolution photography of selected science sites, and S-158 strip photography for two revolutions.

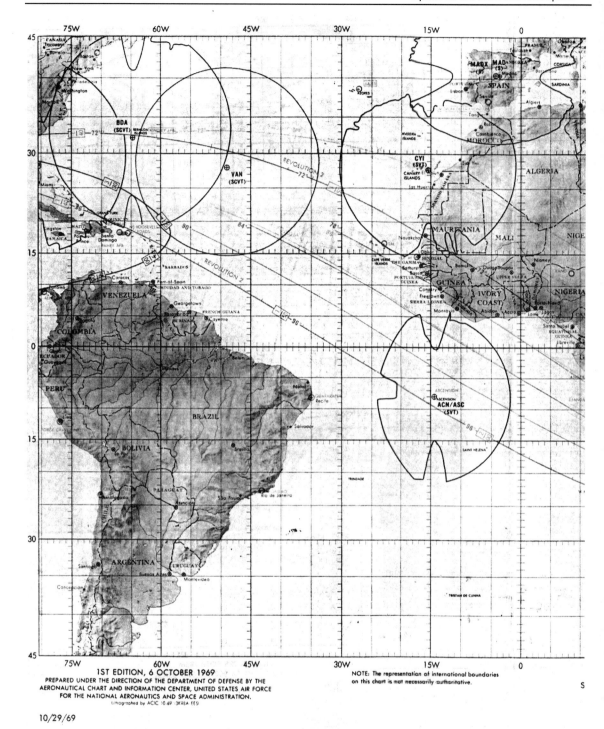

1ST EDITION, 6 OCTOBER 1969
PREPARED UNDER THE DIRECTION OF THE DEPARTMENT OF DEFENSE BY THE
AERONAUTICAL CHART AND INFORMATION CENTER, UNITED STATES AIR FORCE
FOR THE NATIONAL AERONAUTICS AND SPACE ADMINISTRATION.
Lithographed by ACIC 10.69 :3KREA EESi

NOTE: The representation of international boundaries
on this chart is not necessarily authoritative.

S

10/29/69

TEI will then be performed and the nominal mission timeline will be reentered.

Contingency: DPS No-Go for burn (DPS is only failure).

Alternate Mission: The Commander and Lunar Module Pilot will return to the CSM and the LM will be jettisoned. A CSM plane change will be initiated at approximately 116 hours GET which will move the line of nodes backward allowing photographic and landmark tracking of Apollo science sites. During the CSM coast in this orbit, the crew will obtain coverage of eight sites. TEI will be performed on the 41st revolution.

Contingency: LM No-Go for undocking (system failure, not connected with DPS, is discovered during LM

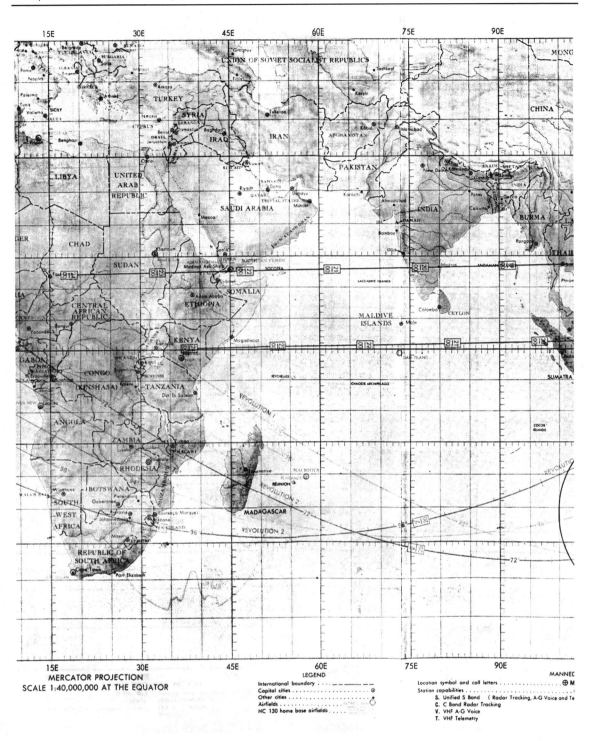

MERCATOR PROJECTION
SCALE 1:40,000,000 AT THE EQUATOR

LEGEND

International boundary . . . . ——— ——
Capital cities . . . . ⊛
Other cities . . . . ○
Airfields . . . .
HC 130 home base airfields . . . . .

Location symbol and call letters . . . . ⊕ M
Station capabilities . . . .
    S. Unified S Band   (Radar Tracking, A-G Voice and Te
    C. C Band Radar Tracking
    V. VHF A-G Voice
    T. VHF Telemetry

checkout).

Alternate Mission: A DPS plane change will be performed. The LM will be jettisoned and landmark tracking and photography of the Apollo science sites will be started. A plane change by the CSM will be initiated at approximately 136 hours GET which will move the line of nodes westward and will allow additional photographic and landmark tracking. This alternate mission will cover ten sites and TEI will be performed on the 40th revolution.

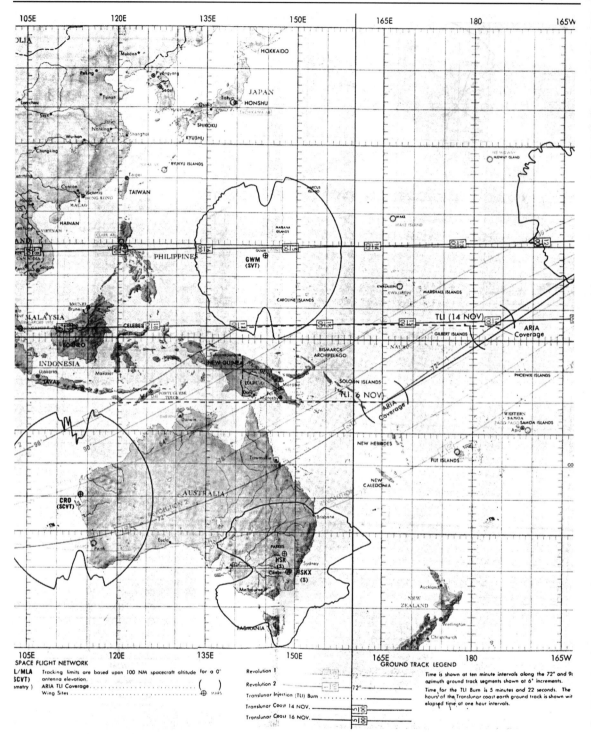

MISSION SUPPORT

GENERAL

Mission support is provided by the Launch Control Center (LCC), the Mission Control Center (MCC), the Manned Space Flight Network (MSFN), and the recovery forces. The LCC is essentially concerned with pre launch checkout, countdown, and with launching the SV; while the MCC, located at Houston, Texas, provides centralized mission control from tower clear through recovery. The MCC functions within the framework of a Communications, Command, and Telemetry System (CCATS); Real-Time Computer Complex (RTCC); Voice Communications System; Display/control System; and a Mission Operations Control Room (MOCR)

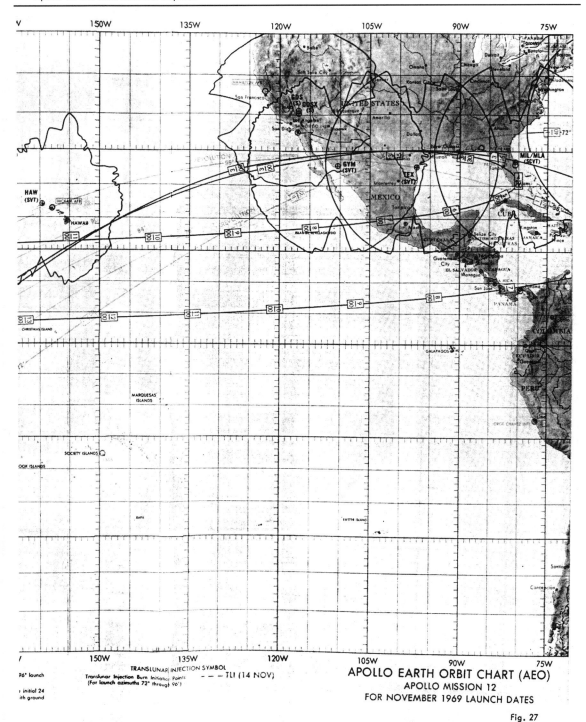

TRANSLUNAR INJECTION SYMBOL
Translunar Injection Burn Initiation Points
(For launch azimuths 72° through 96°)   – – – TLI (14 NOV)

APOLLO EARTH ORBIT CHART (AEO)
APOLLO MISSION 12
FOR NOVEMBER 1969 LAUNCH DATES

Fig. 27

supported by Staff Support Rooms (SSR's). These systems allow the flight control personnel to remain in contact with the spacecraft, receive telemetry and operational data which can be processed by the CCATS and RTCC for verification of a safe mission, or compute alternatives. The MOCR and SSR's are staffed with specialists in all aspects of the mission who provide the Mission Director and Flight Director with real-time evaluation of mission progress.

MANNED SPACE FLIGHT NETWORK

The MSFN is a worldwide communications and tracking network which is controlled by the MCC during Apollo missions (Table 6). The network is composed of fixed stations (Figure 27) and is supplemented by mobile stations. Figure 28 depicts communications during lunar surface operations.

The functions of these stations are to provide tracking, telemetry, updata, and voice communications both on an uplink to the spacecraft and on a downlink to the MCC. Connection between these many MSFN stations and the MCC is provided by NASA Communications Network. More detail on mission support is in the MOR Supplement.

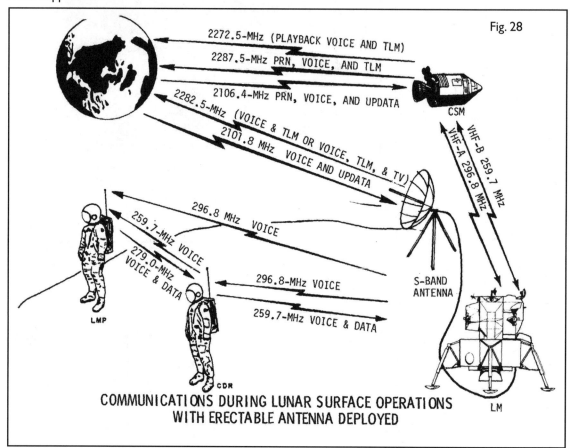

Fig. 28

COMMUNICATIONS DURING LUNAR SURFACE OPERATIONS
WITH ERECTABLE ANTENNA DEPLOYED

RECOVERY SUPPORT

GENERAL

The Apollo 12 flight crew and Command Module (CM) will be recovered as soon as possible after landing, while observing the constraints required to maintain biological isolation of the flight crew, CM, and materials removed from the CM. After locating the CM, first consideration will be given to determining the condition of the astronauts and to providing first-level medical aid when required. Unlike previous spacecraft, the Apollo 12 CM will not deploy the sea dye into the water after landing. The sea dye container and swimmer interphone connector are permanently attached to the upper deck of the CM. If a sea dye marker is requested by the recovery forces, the flight crew will deploy a tethered container of dye through the side hatch of the CM. The container will emit a yellow-green streak in the wake of the CM for approximately 1 hour. The crew has two markers that may be deployed. If the Apollo swimmer radio fails and it becomes necessary to use the interphone to communicate with the crew, the swimmer will have to climb to the top of the CM to reach the interphone connector.

The second consideration will be recovery of the astronauts and CM. Retrieval of the CM main parachutes, apex cover, and drogue parachutes, in that order, is highly desirable if feasible and practical. Special clothing, procedures, and the Mobile Quarantine Facility (MQF) will be used to provide biological isolation of the astronauts and CM. The lunar sample rocks will also be isolated for return to the Manned Spacecraft Center.

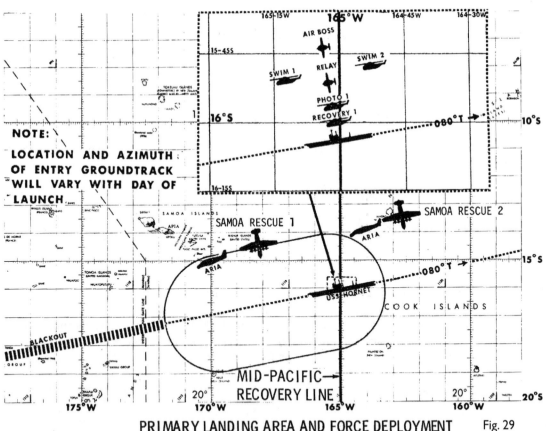

PRIMARY LANDING AREA AND FORCE DEPLOYMENT   Fig. 29

PRIMARY LANDING AREA

The primary landing area, shown in Figure 29, is that area in which the CM will land following circumlunar or lunar orbital trajectories that are targeted to the mid-Pacific recovery line. The target point will normally be 1250 nautical miles (NM) downrange of the entry point (400,000 feet altitude). If the entry range is increased to avoid bad weather, the area moves along with the target point and contains all the high probability landing points as long as the entry range does not exceed 2000 NM.

Recovery forces assigned to the primary landing area are:

* The USS HORNET will be on station at the end-of-mission target point.

* Four SARAH-equipped helicopters, two carrying swimmer teams to conduct electronic search, will be provided. At least one of the swimmers on each team will be equipped with an underwater (Calypso) 35mm camera. Station assignments for these helicopters are:

* One helicopter will be stationed 10 NM uprange from the target point and 15 NM north of the CM ground track.

* One helicopter will be stationed 10 NM downrange from the target point and 15 NM north of the CM ground track.

* One helicopter will be provided for astronaut recovery in the vicinity of the USS HORNET.

* One helicopter carrying photographers as designated by the NASA Recovery Team Leader will be stationed in the vicinity of the USS HORNET.

* One aircraft will fly overhead of the primary recovery ship to function as on-scene commander.

* One aircraft will be on station in the vicinity of the USS HORNET to function as on-scene relay of the recovery commentary.

* Two HC-130 aircraft, each with operational AN/ARD-17 (Cook Tracker), three-man pararescue team, and complete Apollo recovery equipment, will be stationed 100 NM north of the CM ground track. One will be stationed 165 NM uprange, the other 165 NM downrange from the target point.

* Prior to CM entry, two EC-135 Apollo Range Instrumentation Aircraft will be on station near the primary landing area for network support.

The recovery forces will provide the following access and retrieval times:

* A maximum access time of 2 hours to any point in the area.

* A maximum crew retrieval time of 16 hours to any point in the area.

* A maximum CM retrieval time of 24 hours to any point in the area.

Recovery forces will also be provided for support of the launch abort landing area, the secondary landing area, and the contingency landing area. The secondary landing area and the contingency landing area would be used for landing from the earth parking orbit and following aborts during the deep space phase of the mission.

CONFIGURATION DIFFERENCES

SPACE VEHICLE

REMARKS

Command/Service Module (CSM-108)

* Incorporate S-158 Experiment and window modification.

Changes side hatch window pane and adds camera equipment for lunar mission multispectral photography.

* Suppressed Reaction Control System (RCS) engine arc.

Protects Guidance and Navigation System from electromagnetic interference produced by RCS heater switching cycle transients.

* Added Inertial Measurement Unit (IMU) power switch guard.

Prevents loss of IMU due to inadvertent turn off of switch.

* Modified stowage.

Provides for return of Surveyor III samples and increased lunar surface samples.

Lunar Module (LM-6) (Ascent Stage)

* Modified Display and Keyboard Assembly (DSKY) table and support.

Enhances one-handed operation by the flight crew to actuate and release the DSKY table from the stowed to operating position.

* Redesigned ascent stage propellant tanks to all welded configuration.

Eliminates a leak source for propellants and alleviates vehicle weight.

* Added stowable hammocks and blankets.

Increases crew comfort.

* Deleted bacteria filter from forward hatch valve.

Eliminates a potential cause of reduced cabin venting.

* Modified stowage.

Provides for return of Surveyor III samples and increased lunar surface samples.

Lunar Module (LM-6) (Descent Stage)

* Reduced landing gear and plume
deflector thermal insulation.

Reduces vehicle weight by approximately
3.6 lb.

* Modified extravehicular activity
(EVA) equipment stowage.

Provides for current mission requirements.

* Replaced Early Apollo Scientific Experiments Package
(EASEP) with Apollo Lunar Surface Experiments Package
(ALSEP)/Radioisotope Thermoelectric Generator (RTG).

Provides for current mission requirements.

Spacecraft-LM Adapter (SLA-15)

* (No significant differences.)

LAUNCH VEHICLE

Instrument Unit (S-IU-507)

* Added underwater location devices. Increases recovery potential.

S-IVB Stage (SA-507)

Changed the telemetry system for
the S-IVB stage of vehicle SA-507
to consist of one SSB/FM link and
one PCM/DDAS link - SA-506
consisted of one PCM/DDAS link.

Provides for 12 acoustic and 3 vibration
measurements to S-IVB 507 which necessitates
the use of a Saturn MSFC-designed single
sideband/FM system similar to those used
on research and development flights.

S-II Stage (S-11-507)

* (No significant differences.)

S-IC Stage (SA-507)

* (No significant differences.)

MANNED SPACE FLIGHT NETWORK

C-Band Radar

* Deleted PAFB, GBI, GTI, ANT, ASC, PRE, TAN, HAW, CAL.

Eliminates unnecessary duplication of unified S-band.

Unified S-Band

* Deleted GBM, ANG.

Eliminates unnecessary coverage beyond
96° launch azimuth.

* Added pulse modulation capability to PARKES.

Required to support LM descent.

* Added capability to handle LM color TV.

Provides for color TV transmission.

VHF Telemetry

* Deleted GBM, ANG, TAN.

Eliminates unnecessary coverage beyond 96° launch azimuth.

A/G Voice (VHF)

*Deleted GBM, ANG, CAL (GDS
being added), TAN.

Eliminates unnecessary coverage beyond
96° launch azimuth.

Instrumentation Ships and Aircraft

* Deleted USNS REDSTONE,
MERCURY, and HUNTSVILLE.

Eliminates unnecessary real-time coverage
of translunar injection (TLI) and recovery.

* Deleted four Apollo Range Instrumentation Aircraft.

Eliminates unnecessary real-time coverage of TLI.

FLIGHT CREW

FLIGHT CREW ASSIGNMENTS

Prime Crew (Figure 30)

Commander (CDR) - Charles Conrad, Jr. (Commander, USN) Command Module Pilot (CMP) - Richard F. Gordon, Jr. (Commander, USN) Lunar Module Pilot (LMP) - Alan L. Bean (Commander, USN)

Fig. 30

Backup Crew (Figure 31)

Commander (CDR) - David R. Scott (Colonel, USAF) Command Module Pilot (CMP) - Alfred Merrill Worden (Major, USAF) Lunar Module Pilot (LMP) - James Benson Irwin (Lieutenant Colonel, USAF)

The backup crew follows closely the training schedule for the prime crew and functions in three significant categories. One, they receive nearly complete mission training which becomes a valuable foundation for later assignments as a prime crew. Two, should the prime crew become unavailable, they are prepared to fly as prime crew up until the last few weeks prior to launch. Three, they are fully informed assistants who help the prime crew organize the mission and check out the hardware.

During the final weeks before launch, the flight hardware and software, ground hardware and software, and flight crew and ground crews work as an integrated team to perform ground simulations and other tests of the upcoming mission. It is necessary that the flight crew that will conduct the mission take part in these activities, which are not repeated for the benefit of the backup crew. To do so would add an additional costly and time consuming period to the pre launch schedule, which for a lunar mission would require rescheduling for a later lunar launch window.

Fig. 31

PRIME CREW DATA

Commander

NAME: Charles Conrad, Jr. (Commander, USN)

EDUCATION: Bachelor of Science degree in Aeronautical Engineering from Princeton University in 1953; Honorary Master of Arts degree from Princeton University in 1966.

EXPERIENCE: Commander Conrad was selected as an astronaut by NASA in September 1962.

APOLLO: Conrad served as the backup Commander for the Apollo 9 Mission.

GEMINI: On 12 September 1966, Conrad occupied the Command Pilot seat for the 3-day, 44 revolution Gemini 11 Mission. Highlights of the flight included orbital maneuvers to rendezvous and dock in less than one orbit with a previously launched Agena, the retrieval of a nuclear emulsion experiment package during the first extravehicular activity, and the successful completion of the first tethered station keeping exercise, in which artificial gravity was produced.

In August 1965, he served as pilot on the 8-day Gemini 5 Mission. He and Command Pilot Gordon Cooper were launched into orbit on 21 August and proceeded to establish a new space endurance record by which the U.S. took over the lead in man-hours in space.

OTHER: Conrad entered the Navy following his graduation from Princeton University and became a naval aviator. He attended the Navy Test Pilot School at Patuxent River, Maryland, and was then assigned as a project test pilot in the armaments test division there. He also served at Patuxent as a flight instructor and performance engineer at the Test Pilot School .

Command Module Pilot

NAME: Richard F. Gordon, Jr. (Commander, USN)

EDUCATION: Bachelor of Science degree in Chemistry from the University of Washington in 1951.

EXPERIENCE: Commander Gordon was one of the third group of astronauts named by NASA in October 1963.

APOLLO: Gordon served as backup Command Module Pilot for the Apollo 9 Mission.

GEMINI: On 12 September 1966, he served as Pilot for the 3-day Gemini 11 Mission, on which rendezvous with an Agena was achieved in less than one orbit. He executed docking maneuvers with the previously launched Agena and performed two periods of extravehicular activity which included attaching a tether to the Agena and retrieving a nuclear emulsion experiment package. Another highlight of the mission was the successful completion of the first closed-loop controlled entry.

OTHER: Commander Gordon received his wings as a naval aviator in 1953. He then attended all-weather Flight School and jet transitional training and was subsequently assigned to an all-weather fighter squadron at the Naval Air Station at Jacksonville, Florida.

In 1957, he attended the Navy's Test Pilot School at Patuxent River, Maryland, and served as a flight test pilot until 1960. During this tour of duty he did flight test work on the F8U Crusader, F11F Tigercat, FJ Fury, and A4D Skyhawk and was the first project test pilot for the F4H Phantom 11.

He served as a flight instructor in the F4H with Fighter Squadron 121 at the Miramar, California, Naval Air Station and was also flight safety officer, assistant operations officer, and ground training officer for Fighter

Squadron 96 at Miramar.

He was also a student at the U.S. Naval Postgraduate School at Monterey, California.

Lunar Module Pilot

NAME: Alan L. Bean (Commander, USN)

EDUCATION: Bachelor of Science degree in Aeronautical Engineering from the University of Texas in 1955.

EXPERIENCE: Commander Bean was one of the third group of astronauts selected by NASA in October 1963.

APOLLO: Bean served as backup Lunar Module Pilot for the Apollo 9 Mission.

GEMINI: Bean served as backup Command Pilot for the Gemini 10 Mission.

OTHER: Bean, a Navy ROTC student at Texas, was commissioned in 1955 upon graduation from the University. After completing his flight training, he was assigned to Attack Squadron 44 at the Naval Air Station in Jacksonville, Florida, for 4 years. He then attended the Navy Test Pilot School at Patuxent River, Maryland, and was assigned as a test pilot at the Naval Air Test Center, Patuxent River. Commander Bean participated in the trials of both the A5A and the A4E jet attack airplanes. He then attended the Aviation Safety School at the University of Southern California and was next assigned to Attack Squadron 172 at Cecil Field, Florida.

BACKUP CREW DATA

Commander

NAME: David R. Scott (Colonel, USAF)

EDUCATION:. Bachelor of Science degree from the United States Military Academy; degrees of Master of Science in Aeronautics and Astronautics and Engineer of Aeronautics and Astronautics From the Massachusetts Institute of Technology.

EXPERIENCE: Colonel Scott was one of the third group of astronauts selected by NASA in October 1963.

APOLLO: Scott served as Command Module Pilot for Apollo 9, 3-13 March 1969. The 10-day flight encompassed completion of the first comprehensive earth-orbital qualification and verification tests of a "fully configured Apollo spacecraft" and provided vital information previously not available on the operational performance, stability, and reliability of Lunar Module propulsion and life support systems.

GEMINI: On 16 March 1966, he and Command Pilot Neil Armstrong were launched into space on the Gemini 8 Mission - a flight originally scheduled to last 3 days but terminated early due to a malfunctioning spacecraft thruster. The crew performed the first successful docking of two vehicles in space and demonstrated great piloting skill in overcoming the thruster problem and bringing the spacecraft to a safe landing.

OTHER: Scott graduated fifth in a class of 633 at West Point and subsequently chose an Air Force career. He completed pilot training at Webb Air Force Base, Texas, in 1955 and then reported for gunnery training at Laughlin Air Force Base, Texas, and Luke Air Force Base, Arizona.

He was assigned to the 32nd Tactical Fighter Squadron at Soesterberg Air Base (RNLAF), Netherlands, from April 1956 to July 1960. He then returned to the U.S. and completed work on his masters degree at MIT. His thesis at MIT concerned interplanetary navigation.

After completing his studies at MIT in June 1962, he attended the Air Force Experimental Test Pilot School and then the Aerospace Research School.

Command Module Pilot

NAME: Alfred Merrill Worden (Major, USAF)

EDUCATION: Bachelor of Military Science from the United States Military Academy in 1955; Master of Science degrees in Astronautical/Aeronautical Engineering and Instrumentation Engineering from the University of Michigan in 1963.

EXPERIENCE: Major Worden was one of the 19 astronauts named by NASA in April 1966.

APOLLO: Worden served as a member of the astronaut support crew for the Apollo 9 Mission.

OTHER: Major Worden received flight training at Moore Air Base, Texas; Laredo Air Force Base, Texas; and Tyndall Air Force Base, Florida.

Prior to his arrival for duty at the Manned Spacecraft Center, he served as an instructor at the Aerospace Research Pilot's School, from which he graduated in 1965. He is also a graduate of the Empire Test Pilot's School in Farnborough. England, and completed his training there in February 1965.

He attended Randolph Air Force Base Instrument Pilots Instructor School in 1963 and served as a Pilot and Armament Officer from March 1957 to May 1961 with the 95th Fighter Interceptor Squadron at Andrews Air Force Base, Maryland.

Lunar Module Pilot

NAME: James Benson Irwin (Lieutenant Colonel, USAF)

EDUCATION: Bachelor of Science degree in Naval Sciences from the United States Naval Academy in 1951; Master of Science degrees in Aeronautical Engineering and Instrumentation Engineering from the University of Michigan in 1957.

EXPERIENCE: Lt. Colonel Irwin was one of the 19 Astronauts selected by NASA in April 1966.

APOLLO: Irwin was crew Commander of Lunar Module LTA-8; this vehicle finished the first series of thermal vacuum tests on 1 June 1968. He also served as a member of the support crew for Apollo 10.

OTHER: Irwin was commissioned in the Air Force on graduation from the Naval Academy in 1951 and received his flight training at Hondo Air Base, Texas, and Reese Air Force Base, Texas.

He also served with the F-12 Test Force at Edwards Air Force Base, California, and with the AIM 47 Project Office at Wright-Patterson Air Force Base, Ohio.

MISSION MANAGEMENT RESPONSIBILITY

Title	Name	Organization
Director, Apollo Program	Dr. Rocco A. Petrone	NASA/OMSF
Director, Mission Operations	Maj. Gen. John D. Stevenson (Ret)	NASA/OMSF
Saturn Program Manager	Mr. Roy E. Godfrey	NASA/MSFC
Apollo Spacecraft Program Manager	Col. James A. McDivitt	NASA/MSC
Apollo Program Manager KSC	Mr. Edward R. Mathews	NASA/KSC
Mission Director	Capt. Chester M. Lee (Ret)	NASA/OMSF
Assistant Mission Director	Col. Thomas H. McMullen	NASA/OMSF
Director of Launch Operations	Mr. Walter J. Kapryan	NASA/KSC
Director of Flight Operations	Dr. Christopher C. Kraft	NASA/MSC
Launch Operations Manager	Mr. Paul C. Donnelly	NASA/KSC
Flight Directors	Mr. Gerald D. Griffin Mr. M. P. Frank Mr. Glynn S. Lunney Mr. Clifford E. Charlesworth	NASA/MSC
Spacecraft Commander (Prime)	Cdr. Charles Conrad, Jr.	NASA/MSC
Spacecraft Commander (Backup)	Col. David R. Scott	NASA/MSC

PROGRAM MANAGEMENT

NASA HEADQUARTERS

Office of Manned Space Flight
Manned Spacecraft Center
Marshall Space Flight Center
Kennedy Space Center

LAUNCH VEHICLE	SPACECRAFT	TRACKING AND DATA ACQUISITION
Marshall Space Flight Center The Boeing Co. (S-IC) North American Rockwell Corp. (S-II) McDonnell Douglas Corp. (S-IVB) IBM Corp. (IU)	Manned Spacecraft Center North American Rockwell (LES, CSM, SLA) Grumman Aerospace Corp. (LM)	Kennedy Space Center Goddard Space Flight Center Department of Defense MSFN

ABBREVIATIONS AND ACRONYMS

ACN	Ascension Island
A/G	Air To Ground
AGS	Abort Guidance System
ALSEP	Apollo Lunar Surface Experiments Package
ANG	Antigua Island (MSFN)
ANT	Antigua Island (DOD)
APS	Ascent Propulsion System (LM)
APS	Auxiliary Propulsion System (S-IVB)
ARIA	Apollo Range Instrumentation Aircraft
AS	Apollo/Saturn
ASC	Ascension Island
BDA	Bermuda
BIG	Biological Isolation Garment
CAL	Point Arguello, California
CCATS	Communications, Command, and Telemetry System
CD	Countdown
CDH	Constant Delta Height
CDR	Commander
CM	Command Module
CMD	Command
CMP	Command Module Pilot
CNV	Cape Canaveral
CRO	Carnarvon
CSI	Concentric Sequence Initiation
CSM	Command/Service Module
CYI	Grand Canary Island
DDAS	Digital Data Acquisition System
DOD	Department of Defense
DOI	Descent Orbit Insertion
DPS	Descent Propulsion System
DSKY	Display and Keyboard Assembly
EASEP	Early Apollo Scientific Experiments Package
EI	Entry Interface
EMU	Extravehicular Mobility Unit
EPO	Earth Parking Orbit
EST	Eastern Standard Time
ETB	Equipment Transfer Bag
EVA	Extravehicular Activity
FM	Frequency Modulation
fps	Feet Per Second
FT	Feet
FTP	Full Throttle Position
GBI	Grand Bahama Island (USAF)
GBM	Grand Bahama Island (NASA)
GDS	Goldstone, California
GET	Ground Elapsed Time
GTI	Grand Turk Island (NASA)
GTK	Grand Turk Island (DOD)
GYM	Guaymas, Mexico
HAW	Kauai, Hawaii
HR	Hour
HTC	Hand Tool Carrier
IMU	Inertial Measurement Unit

IU	Instrument Unit
IVT	Intravehicular Transfer
KSC	Kennedy Space Center
LCC	Launch Control Center
LEC	Lunar Equipment Conveyor
LES	Launch Escape System
LH2	Liquid Hydrogen
LiOH	Lithium Hydroxide
LM	Lunar Module
LMP	Lunar Module Pilot
LOI	Lunar Orbit Insertion
LOX	Liquid Oxygen
LPO	Lunar Parking Orbit
LR	Landing Radar
LRL	Lunar Receiving Laboratory
LV	Launch Vehicle
m	Meter
mm	Millimeter
MAD	Madrid
MAX	Maximum
MCC	Midcourse Correction
MCC	Mission Control Center
MESA	Modularized Equipment Stowage Assembly
MHz	Megahertz
MIL	Merritt Island (NASA)
MLA	Merritt Island (DOD)
MIN	Minute
MOCR	Mission Operations Control Room
MOR	Mission Operations Report
MQF	Mobile Quarantine Facility
MSC	Manned Spacecraft Center
MSFC	Marshall Space Flight Center
MSFN	Manned Space Flight Network
NASA	National Aeronautics and Space Administration
NASCOM	NASA Communications Network
NM	Nautical Mile
OMSF	Office of Manned Space Flight
OPS	Oxygen Purge System
PAFB	Patrick Air Force Base
PAT	Patrick AFB
PCM	Pulse Code Modulation
PDI	Powered Descent Initiation
PGNCS	Primary Guidance, Navigation, and Control System
PLSS	Portable Life Support System
PRE	Pretoria
PRN	Pseudorandom Noise
psi	Pounds Per Square Inch
QUAD	Quadrant
RCS	Reaction Control System
RLS	Radius Landing Site
RNLAF	Royal Netherlands Air Force
RTCC	Real-Time Computer Complex
RTG	Radioisotope Thermoelectric Generator
S/C	Spacecraft
SEA	Sun Elevation Angle

S-IC	Saturn V First Stage
S-II	Saturn V Second Stage
S-IVB	Saturn V Third Stage
SLA	Spacecraft-LM Adapter
SM	Service Module
SPAN	Solar Particle Alert Network
SPS	Service Propulsion System
SRC	Sample Return Container
SSB	Single Side Band
SSR	Staff Support Room
SV	Space Vehicle
SWC	Solar Wind Composition
TAN	Tananarive, Malagasy Republic
TEI	Trans-earth Injection
TEX	Corpus Christi, Texas
TFI	Time From Ignition
TLM	Telemetry
TLI	Translunar Injection
TPF	Terminal Phase Finalization
TPI	Terminal Phase Initiation
TRAJ	Trajectory
T-time	Countdown Time (referenced to liftoff time)
TTY	Teletype
TV	Television
USB	Unified S-band
USN	United States Navy
USAF	United States Air Force
VAN	Vanguard
VHF	Very High Frequency
Delta-V	Differential Velocity

Post Launch
Mission Operation Report

No. M-932-69-12

TO: A/Administrator
25 November 1969

FROM: MA/Apollo Program Director

SUBJECT: Apollo 12 Mission (AS-507)

Post Launch Mission Operation Report #1

The Apollo 12 Mission was successfully launched from the Kennedy Space Center on Friday, 14 November 1969 and was completed as planned, with recovery of the spacecraft and crew in the Pacific Ocean recovery area on Monday, 24 November 1969. Initial review of the flight indicates that all mission objectives were attained. Further detailed analysis of all data is continuing and appropriate refined results of the mission will be reported in the Manned Space Flight Centers' technical reports.

Attached is the Mission Director's Summary Report for Apollo 12 which is hereby submitted as Post Launch Mission Operation Report #1. Also attached are the NASA OMSF Primary Mission Objectives for Apollo 12. I recommend that the Apollo 12 Mission be adjudged as having achieved all the established Primary Objectives and be considered a success.

Rocco Petrone

APPROVAL:

George E. Mueller
Associate Administrator for
Manned Space Flight

NASA OMSF PRIMARY MISSION OBJECTIVES FOR APOLLO 12
PRIMARY OBJECTIVES

Perform selenological inspection, survey, and sampling in a mare area.

Deploy and activate an Apollo Lunar Surface Experiments Package (ALSEP).

Develop techniques for a point landing capability.

Develop man's capability to work in the lunar environment.

Obtain photographs of candidate exploration sites.

Rocco A. Petrone   George E. Mueller
Apollo Program Director   Associate Administrator for
                                         Manned Space Flight

Date: 31 October 1969   Date: 11-3-69

RESULTS OF APOLLO 12 MISSION

Based upon a review of the assessed performance of Apollo 12, launched 14 November 1969 and completed 24 November 1969, this mission is adjudged a success in accordance with the objectives stated above.

Rocco A. Petrone   George E. Mueller
Apollo Program Director   Associate Administrator for
                           Manned Space Flight

Date: 25 November 1969   Date: 25 November 1969

NATIONAL AERONAUTICS AND SPACE ADMINISTRATION
WASHINGTON, D.C. 20546

IN REPLY REFER TO: MAO

25 November 1969

TO: Distribution
FROM: MA/Apollo Mission Director
SUBJECT: Mission Director's Summary Report, Apollo 12

INTRODUCTION

The Apollo 12 Mission was planned as a lunar landing mission to: perform selenological inspection, survey, and sampling in a mare area; deploy and activate an Apollo Lunar Surface Experiments Package (ALSEP); develop techniques for a point landing capability; develop man's capability to work in the lunar environment; and obtain photographs of candidate exploration sites. Flight crew members were: Commander (CDR), Cdr. Charles Conrad, Jr.; Command Module Pilot (CMP), Cdr. Richard F. Gordon, Jr.; Lunar Module Pilot (LMP), Cdr. Alan L. Bean. Significant detailed mission information is contained in Tables I through 11. Initial review of the flight indicates that all mission objectives were attained (Reference Table 1). Table 2 lists Apollo 12 mission achievements.

PRELAUNCH

An unscheduled 6-hour hold occurred at T-17 hours (spacecraft cryogenic loading) in order to replace Service Module (SM) liquid hydrogen tank No. 2 which had been leaking. The weather conditions at launch were: peak ground winds of 14 knots, light rain showers, broken clouds at 800 feet, and overcast at 10,000 feet with tops at about 21,000 feet.

LAUNCH AND EARTH PARKING ORBIT

The Apollo 12 space vehicle was successfully launched on schedule from Kennedy Space Center, Florida, at 11:22 a.m. EST on 14 November 1969. All launch vehicle stages performed satisfactorily, inserting the S-IVB/IU/LM/CSM combination into an earth parking orbit with an apogee of 102.5 nautical miles (NM) and a perigee of 99.9 NM (103 NM circular planned). All launch vehicle systems operated satisfactorily except for two minor off-nominal conditions which were noted in the launch vehicle digital computer. During the ascent (36.5 to 52 seconds ground elapsed time (GET)) a number of spacecraft electrical transients also occurred. The tentative conclusion is that the cause of these events was an electrical potential discharge from the clouds through the space vehicle to the ground.

After orbital insertion, launch vehicle and spacecraft systems were verified, preparations were made for translunar injection (TLI) as planned, and the second S-IVB burn was initiated on schedule (Reference Tables 3, 4, and 5). All major systems operated satisfactorily and all end conditions were nominal for a free-return circumlunar trajectory. The prelaunch planned height of closest approach of the spacecraft after the TLI maneuver was 1851 NM prior to the second midcourse correction, MCC-2, as shown in Table 5. The actual height of closest approach, after TLI and prior to MCC-2, was estimated to be 457 NM, the spacecraft still being injected on a free-return trajectory. This difference appears to be due to a state vector error in the Saturn Instrument Unit (IU). The error was known before TLI, but because of time limitations, the decision was made to ignore it and not change the TLI targeting.

TRANSLUNAR COAST

The Command/Service Module (CSM) separated from the LM/IU/S-IVB at about 3:18 (hr:min) GET. Onboard television was initiated shortly thereafter and clearly showed docking with the Lunar Module (LM) at 3:27 GET. Ejection of the CSM/LM was successfully accomplished at about 4:14 GET and an S-IVB Auxiliary

Propulsion System (APS) evasive maneuver was performed (and observed on television) at 4:27 GET. All launch vehicle safing activities were performed as scheduled.

The S-IVB slingshot maneuver was initiated on schedule. The total APS burn time was 570 seconds of which 270 seconds were due to a commanded burn. Due to the some IU state vector errors that affected the TLI result, the slingshot maneuver did not achieve the desired heliocentric orbit but rather a geocentric orbit with the following parameters: period - 39 to 45 days; apogee - 448,000 to 487,000 NM; perigee - 81,000 to 95,000 NM. The S-IVB/IU closest approach to the moon was 3091 NM at 85:48 GET.

To insure that the electrical transients noted in the CSM during launch had not affected the LM systems, the CDR and LMP entered the LM earlier than planned, at about 7:20 GET, to perform some of the housekeeping and systems checks. All checks indicated that the LM systems were satisfactory.

The TLI maneuver parameters were accomplished such that the CSM/LM were on an acceptable free-return trajectory and MCC-1, scheduled for 11:53 GET, was not required.

MCC-2 was performed as planned at 30:53 GET and resulted in placing the spacecraft on the desired hybrid, non-free-return circumlunar trajectory with a closest approach of 60 NM. All SM Service Propulsion System (SPS) burn parameters were normal.

Good quality television coverage of the preparations and performance of MCC-2 was received for 47 minutes. The accuracy of MCC-2 was such that neither MCC-3 (scheduled for 61:31 GET) nor MCC-4 (78:31 GET) was required.

The CDR and LMP began their intravehicular transfer to the LM during translunar coast about 1/2 hour earlier than planned in order to obtain full television coverage through the Goldstone tracking station. The 56-minute transmission, beginning at 62:52 GET, showed excellent color pictures of the Command Module (CM), intravehicular transfer, the LM interior, and brief shots of the earth and moon. The television clearly showed numerous linear streaks in CM window number 1 as previously reported. The crewmen completed their activities in the LM in under 40 minutes.

LUNAR ORBIT

Lunar orbit insertion (LOI) was performed in two separate maneuvers, LOI-1 and LOI-2, using the SPS. LOI-1, initiated at 83:25 GET, placed the spacecraft in a 168.8 by 62.6-NM elliptical orbit. LOI-2, initiated two revolutions later at 87:49 GET, placed the spacecraft in a near-circular orbit of 66.1 by 54.3 NM. Table 6 summarizes the maneuvers performed by the CSM and LM in lunar orbit.

During the first lunar orbit, good quality television coverage of the lunar surface was received for about 33 minutes. The crew provided excellent descriptions of the lunar features while transmitting sharp pictures back to earth. A television broadcast scheduled for 81:30 GET prior to LOI-1 was cancelled due to the sun angle and the glare on the spacecraft windows.

One revolution after LOI-2 the LM crew transferred to the LM and performed various housekeeping chores and communications checks in preparation for lunar descent the following day. The intravehicular transfer and LM activities lasted about 1½ hours, during which time the LM was powered up about 22 minutes.

CSM/LM docking took place at 107:54 GET. Television pictures clearly showed the LM landing gear to be in the deployed position. The CSM separation maneuver was successfully executed as planned at 108:25 GET, using the SM Reaction Control System (RCS). The descent orbit insertion (DOI) maneuver was also successfully executed as planned at 109:24 GET, using the LM Descent Propulsion System (DPS).

DESCENT

The LOI maneuver resulted in a CSM/LM position some 4 to 5 NM north of the expected ground track prior to DOI. This crossrange error was known prior to DOI and was corrected during the powered descent maneuver, which was on time at 110:21 GET. The initial LM roll angle resulted from an initial crossrange distance of 4.9 NM. The LM guidance computer was updated during powered descent to compensate for indications that the trajectory was coming in 4200 feet short of the target point. The initial crossrange distance was continuously reduced throughout the braking phase. At entry into the approach phase the spacecraft's trajectory was very close to nominal. Redesignations were incorporated during the approach phase. The crew took over manual control at about 370 feet, passed over the right side of the target crater, then flew to the left for landing at 110:32:35 GET (01:54:35 a.m. EST, 19 November.) The Commander reported extensive dust obscuring his view during final descent. The actual landing point is estimated to be about 600 feet from the Surveyor III spacecraft. Landing coordinates are 3.036°S, 23.418°W. LM tilt on the surface was 3.5° from the vertical and the LM was pointing on an azimuth of 295° (10° right of the approach path).

During the next CSM revolution, the Commander reported a visual sighting of the CSM orbiting overhead. On the following CSM revolution, the CMP reported sighting the Surveyor III spacecraft as well as the LM northwest of Surveyor III.

LUNAR SURFACE

The first extravehicular activity period (EVA-1) started at 115:11 GET, about 1/2 hour later than planned due to time spent in establishing the location of the landed LM and general preparations for EVA. The Commander egressed from the LM, went partway down the ladder, and deployed the Modularized Equipment Stowage Assembly (MESA) and color television camera. He reported seeing the Surveyor III spacecraft about 600 feet away and also stated that the LM had landed about 25 feet downrange from the lip of a crater. The Commander first touched the lunar surface at 115:22 GET (06:44 a.m. EST, 19 November). His descriptions indicated that the lunar surface was quite soft and loosely packed, causing his boots to dig in as he walked.

The LMP descended to the lunar surface at 115:52 GET (07:14 a.m. EST). Shortly after the television camera was removed from the bracket in the MESA, transmission was lost, and, despite repeated efforts, was not regained for the remainder of the EVA. The lithium hydroxide canisters and contingency sample were transferred to the LM cabin as planned. Deployment of the S-band erectable antenna, the Solar Wind Composition experiment, and the American flag was accomplished as planned.

Except for minor difficulty removing the radioisotope thermoelectric generator fuel element from the cask, the removal of the Apollo Lunar Surface Experiments Package (ALSEP) from the MESA, transport to the site, and deployment was accomplished nominally. The ALSEP deployment site was estimated to be 600 to 700 feet from the LM. Shortly after deployment, the Passive Seismometer transmitted to earth the crewmen's footsteps as they returned to the LM.

A considerable amount of dust was kicked up by the astronauts during ALSEP deployment and some adhered to the instruments. The overall effect will be determined through the long-term measurements of the engineering parameters of the system. The crewmen dusted each other off prior to ingressing the LM. The LM was repressurized, concluding a total EVA time of 3 hours 56 minutes. The first CSM plane change maneuver, LOPC-1, was successfully accomplished as planned at 119:47 GET.

Prior to the second LM egress a plan for the second extravehicular activity period (EVA-2) had been formulated by the astronauts, earth-based geologists, and Mission Control. The EVA-2 traverse (Figure 1) took the crew to the ALSEP deployment site, Head Crater, Bench Crater, Sharp Crater, Halo Crater, Surveyor III, Block Crater, and back to the LM. The astronauts walked between 1500 and 2000 feet from the LM, and the total distance traversed was about 6000 feet. As the astronauts gained experience in walking on the lunar surface their confidence and speed increased significantly, as evidenced throughout EVA-2.

EVA-2 began at 131:33 GET (10:55 p.m. EST, 19 November), 1 hour 40 minutes ahead of schedule, and lasted for 3 hours 49 minutes. The astronauts first cut the cable and stored the inoperative LM TV camera in the Equipment Transfer Bag for return to earth and subsequent failure analysis. The Commander then went to

M-932-69-12

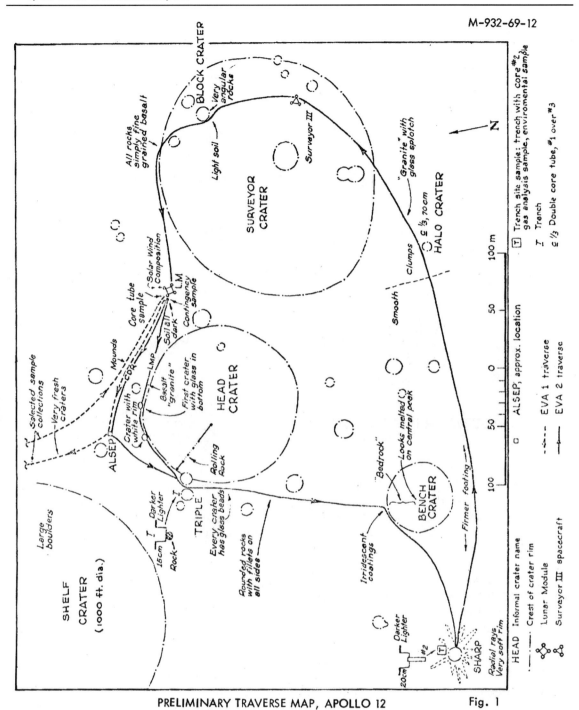

PRELIMINARY TRAVERSE MAP, APOLLO 12   Fig. 1

the ALSEP site to check the leveling of the Lunar Atmosphere Detector (Cold Cathode Ion Gauge). As he approached the instrument, it recorded a higher atmosphere, rising to 10^{-6} torr (one millionth of a millimeter of mercury). This rise is attributed to the outgassing of the astronaut's suit.

Astronaut movement on the lunar surface was recorded on the Passive Seismometer and on the Lunar Surface Magnetometer. In addition, the Commander rolled a grapefruit sized rock down the wall of Head Crater, about 300 to 400 feet from the Passive Seismometer. No significant response was detected on any of the four axes.

During the geological traverse, the crewmen obtained the desired photographic panoramas, stereo

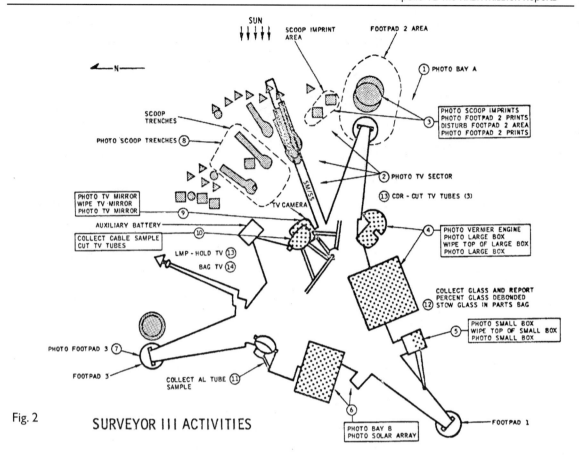

Fig. 2   SURVEYOR III ACTIVITIES

photographs, core samples (2 single, I double), trench (8-inch deep) sample, lunar environment samples and assorted rock, dirt, bedrock, and "molten" samples. They reported seeing fine dust buildup on all sides of larger rocks and that soil color seemed to become lighter as they dug deeper. The Apollo Lunar Surface Close-up Camera was used to take stereo pictures in the vicinity of the LM during the last few minutes of EVA-2.

All Surveyor III activities were accomplished as planned (Figure 2). In addition, the soil scoop was removed and retrieved by the crew. They reported that the Surveyor footpad marks were still visible and that the entire spacecraft had a brown appearance. The glass parts were not broken, only warped slightly on their mountings, and therefore were not retrieved.

Following the geological traverse, the Solar Wind Composition experiment was retrieved and stowed in the Equipment Transfer Bag. Some difficulty was experienced in the retrieval operation. All of the collected samples, parts, and equipment were then transferred into the LM, using the Lunar Equipment Conveyor. The crewmen dusted each other off prior to ingressing the LM. Expendable equipment was jettisoned and the cabin was repressurized.

During the LM. lunar surface stay period the S-158 Lunar Multispectral Photography Experiment was completed by the CMP in the CSM according to plan. In addition, photography of three desirable targets of opportunity was obtained: the Wall of Theophilus and two future Apollo landing sites, Fra Mauro and Descartes. The returned film will be analyzed to aid scientists in planning for future sample collection and in extrapolating known compositions from returned samples to other parts of the moon that will not be visited by man. The CMP reported that, in his judgement, the condition of the CM windows did not degrade the quality of the photographs.

ASCENT AND RENDEZVOUS

Lunar liftoff occurred at 142:04 GET (09:26 a.m. EST, 20 November) concluding a total lunar stay time of 31 hours 31 minutes. A 1.2-second overburn of the LM Ascent Propulsion System resulted in an insertion velocity 32 feet per second greater than planned, placing the LM in a 62 by 9.2-NM orbit. The overburn was caused by an incorrect manual switching sequence, which prevented the automatic shutdown command from shutting off the engine. The crew quickly recognized the discrepancy and manually shut down the engine. Subsequently, they used an RCS trim maneuver to return to an orbit with the planned parameters of 46.3 by 8.8 NM.

The rendezvous sequence of maneuvers occurred as planned. The Commander in the LM reported first visually sighting the CSM at a range of 122 NM. Good quality television was transmitted from the CSM for 24 minutes during the final portions of the rendezvous sequence. Docking was accomplished at 145:36 GET and was clearly seen on television. Intravehicular transfer of the crew and equipment from the LM to the CSM was accomplished without difficulty. LM jettison and CSM separation took place normally at 148:00 GET and 148:05 GET, respectively.

POST-RENDEZVOUS

The ascent stage deorbit retrograde burn was initiated at 149:28 GET and burned slightly longer than planned. This resulted in lunar impact about 36 NM short of the target point (Figure 3). Impact occurred at 149:55:17 GET (05:17:17 p.m. EST, 20 November), about 39 NM southeast of Surveyor III. Coordinates of impact were 3.95°S, 21.17°W as compared to the target point of 3.34°S, 23.42°W. LM weight and velocity at impact were 5254 pounds and 5502 feet per second, respectively. The flight path angle at impact was -3.792 degrees.

Following the completion of LM activities, the crew performed housekeeping functions in the CSM. The second CSM plane change maneuver (LOPC-2) was successfully executed at 159:05 GET. The resultant orbit was as planned for the subsequent lunar reconnaissance photography.

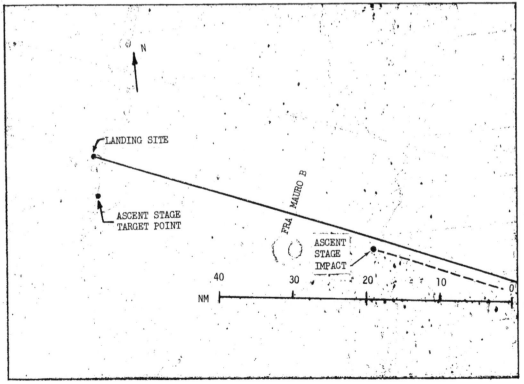

Fig. 3

LANDING AND IMPACT LOCATIONS

High resolution photography, landmark tracking, and stereo strip photography were conducted during this period to conclude the required Flight Plan activities over the sites Fra Mauro, Descartes, and Lalande.

TRANSEARTH AND ENTRY

The transearth injection (TEI) maneuver was successfully performed on time using the SPS. All systems performed normally and initial tracking shortly after the spacecraft emerged from behind the moon indicated that trajectory parameters were nominal.

Good quality television of the receding moon and spacecraft interior was received for about 38 minutes, beginning about 20 minutes after TEI.

MCC-5 was postponed until 188:28 GET, 1 hour later than planned, to allow additional crew rest, and the 2.0-foot per second maneuver was performed nominally. MCC-6, showing a requirement for less than 1 foot per second, was not performed.

Good quality television of the spacecraft interior and a question and answer period with scientists and members of the press began at 224:07 GET and lasted approximately 37 minutes.

The final transearth midcourse correction, MCC-7, was initiated at 241:24 GET resulting in a predicted entry velocity of 36,116 feet per second and a flight path angle of -6.47 degrees.

CM/SM separation occurred at 244:07 GET and entry interface (EI) at 400,000 feet altitude was reached at 244:22 GET. Visual contact with the spacecraft was reported at 244:33 GET. Drogue and main parachutes deployed normally. Landing occurred about 14 minutes after EI at 244:35:25 GET (03:58:25 p.m. EST). The landing point was in the mid-Pacific Ocean, approximately 165°10'W longitude and 15°44'S latitude. The CM landed in the Stable 2 position about 3.5 NM from the prime recovery ship, USS HORNET. Flotation bags were deployed to right the spacecraft to the Stable 1 position at 244:41 GET. The crew reported that they were in good condition.

Weather in the prime recovery area was good; visibility 10 miles, wind 15 knots, and wave height 3 feet with swells about 15 feet.

ASTRONAUT RECOVERY OPERATIONS

Following landing, the recovery helicopter dropped swimmers who installed the flotation collar to the CM. A raft was deployed and attached to the flotation collar. Flight suits and oral/nasal masks were lowered into the raft and passed in to the crew through the spacecraft hatch. The postlanding ventilation fan was turned off, the CM was powered down, and the astronauts egressed. The swimmer closed the CM hatch and then decontaminated all garments, the hatch area, the collar, and the area around the postlanding vent valve.

The helicopter recovered the astronauts. After landing on the recovery carrier, the astronauts and a recovery physician entered the Mobile Quarantine Facility (MQF).

The flight crew, recovery physician, and recovery technician will remain inside the MQF until it is delivered to the Lunar Receiving Laboratory (LRL) in Houston, Texas. This delivery is currently planned to occur on 29 November.

COMMAND MODULE RETRIEVAL OPERATIONS

After flight crew pickup by the helicopter, the CM was retrieved and placed in a dolly aboard the recovery ship. It was then moved to the MQF and mated to the transfer tunnel. From inside the MQF/CM containment envelope, the MQF engineer began post-retrieval procedures (removal of lunar samples, data, equipment, etc.), passing the removed items through the decontamination lock. The CM will remain sealed during RCS deactivation and delivery to the LRL.

The Sample Return Containers (SRC), film, data, etc., will be flown to Pago Pago by fixed-wing aircraft from USS HORNET, and then by separate aircraft to Houston for transport to the LRL. Both of the SRC's should arrive at the LRL on 25 November.

MISSION SCIENTIFIC ACTIVITY

Apollo Lunar Surface Experiments Package

The Apollo Lunar Surface Experiments Package (ALSEP) was deployed on the lunar surface by the Apollo 12 crew on 19 November 1969. The ALSEP deployment site was estimated to be 600 to 700 feet from the LM. Activation took place at 9:21 a.m. EST by "transmitter on" command from the ground. All initial conditions were normal.

The ALSEP is powered by SNAP-27, a radioisotope thermoelectric generator which uses plutonium 238 as a fuel. The initial power output of the generator was 56.74 watts. As the generator warmed up, the power output increased and stabilized at 73.59 watts.

The dust detector cells are operating well. The top and east cells show a slight disagreement that could be caused by tilt of the central station.

The Passive Seismic Experiment (PSE) recorded astronaut operation of the core tube, foot steps, and other crew activities. The seismometer measured the effects of the LM ascent at 9:26 a.m. EST on 20 November 1969, and the subsequent impact of the LM ascent stage at 5:17 p.m. EST. The PSE showed significant response to the impact of the jettisoned ascent stage. The impact was 39 NM (43 statute miles) from the ALSEP. The signal received at the PSE recorded on all three long period axes, but not on the short period vertical, indicating that the wave traveled through the lunar surface layer but was not strong enough to travel through the body of the moon. The signal was low frequency (1 hertz), extremely low velocity, and lasted nearly 1 hour. This long duration was not anticipated and is outside of our experience here on earth.

Although the significance of the event is not now understood, it appears similar to the signals obtained for the Apollo 11 seismic experiment. Since scientists know the amount of energy of the impact, location, time, etc., detailed analyses are expected to provide an explanation of the event.

The Lunar Surface Magnetometer (LSM) was aligned and leveled to 0.2°. The magnetic field measurements indicated that the moon was passing through the earth's bow shock during the first day of operation and subsequently passed through the transition region into the geomagnetic tail during 20, 21, and 22 November 1969. Five calibrations have been performed to date and the results indicate that the instrument is performing as expected. The site survey operation was performed between 5:50 and 6:40 p.m. EST on 22 November 1969. This important operation verified that the LSM measurements are not influenced by any local magnetic field anomalies.

The Solar Wind Spectrometer (SWS) was turned on at 1:40 p.m. EST on 19 November 1969. The dust covers for the seven Faraday cups were blown at 10:25 a.m. EST, 20 November 1969. Plasma ion data characteristic of the earth's "transition region" was observed. These data were consistent with the indications from the magnetometer that ALSEP passed behind the plasma bow shock of the earth near the time of deployment.

The SWS is equipped with a solar cell, located under a very narrow slit that allows sunlight to pass through at a sun angle of 60 degrees. The solar cell operated as planned, and, based on this data, the principal investigator estimates that the Apollo 12 crew leveled this instrument to within 1.2 degrees of actual level. Deployment tolerance was ±5 degrees.

The Lunar Ionosphere Detector (Suprathermal Ion Detector Experiment - SIDE) which measures the lunar ionosphere, was activated at 2:18 p.m. EST on 19 November 1969. The dust cover was inadvertently opened during deployment and was left open. The total ion detector and the mass analyzer both have returned interesting and useful scientific data. The background counting rates indicated outgassing from the interior of the electronics package. This outgassing causes arcing, which does not damage the instrument, but which prevents it from detecting scientific data. The high voltage (3.5 kilovolts) power supply was turned off until after lunar noon, at which time outgassing is expected to be complete. The instrument will then be commanded back on high voltage.

The Lunar Atmosphere Detector (Cold Cathode Ion Gauge - CCIG) which measures the ambient lunar atmosphere, was turned on between the first and second EVA's. During the astronaut sleep period prior to liftoff, the CCIG's high voltage power supply turned itself off. This shut-off is probably due to arcing similar to the condition described for the SIDE. The CCIG high voltage power supply has also been turned off until outgassing is complete.

The primary objective of the CCIG to measure the lunar atmosphere has not yet been achieved. It would not have been achieved by this time even if the experiment were operating perfectly, since there is still considerable contamination in the area of the landing.

During the 14 hours of CCIG data, the indicated pressure dropped at a decreasing rate. When the LM was vented for the second EVA, a slight pressure increase was noticed, and when the Commander walked near the gauge a large pressure increase was noted.

The Solar Wind Composition Experiment is designed to measure abundances and isotopic compositions of noble gases in the solar wind. It consists of a specially prepared aluminum foil that is unrolled on a staff implanted in the lunar surface. The experiment was deployed at 7:35 a.m. EST on 19 November 1969 and a foil exposure time of 18 hours 42 minutes was achieved. The foil is being returned in the documented sample Sample Return Container (SRC).

Lunar Geology

The Apollo 12 landing site is a gently rolling surface that includes several large, subdued craters and many smaller craters with raised rims. It is underlain everywhere by a regolith consisting of very fine-grained, gray, particulate material with admixed glassy beads, clasts, and coatings in other rocks. Blocks from several centimeters to 7 meters across are sparsely strewn in the regolith.

The Apollo 12 landing site is on a mare (Ocean of Storms), At this locality, the mare appears to consist of basalt flows overlain by a regolith that varies in thickness from 1 to 5 meters. The site also lies on a broad ray associated with the large crater Copernicus, 400 km to the north.

Contingency Sample

A contingency sample was acquired early in the first EVA in the immediate vicinity of the LM. This sample includes several rock fragments, fine-grained material, and at least one piece of glassy material.

Selected Sample

After deployment of the ALSEP, individual samples were taken on a traverse that extended to a point approximately 350 meters from the LM to a crater approximately 450 meters in diameter. Photographs were taken of several sample locations. One core tube sample was taken near the LM and SRC No. 1 was then filled with fine-grained material.

Documented Sample

At least 20 well documented samples were collected along the 6000-foot traverse performed during the second EVA. An environmental sample was taken from a shallow trench and sealed in the Special Environmental Sample Container. The environmental sample, core tube sample, and gas analysis sample were taken near Sharp crater, the farthest point from the LM. The two remaining core tubes were joined together and successfully driven in the surface near Halo crater to get a core sample approximately 70 cm long.

Apollo 12 with tower & gantry.

A spectacular shot of Apollo 12 as it leaves
the Vehicle Assembly Building.

Charles "Pete" Conrad veteran
of two Gemini missions and
Commander of Apollo 12.

Alan Bean
Lunar Module Pilot
of Apollo 12.

Preflight crew shot.
Conrad, Gordon and
Bean pose with several
ALSEP experiments (left).

Richard Gordon
Gemini veteran
and Command Module Pilot
of Apollo 12.

Apollo 12 crew heading for the moon. (Above)

November 14th 1969 Saturn V burning off 5.7 million pounds of propellant in nine minutes. (Left and Right)

Dick Gordon inside the CMS. (Right)
Lightning strike is captured on film moments after Apollo 12 has departed Pad 39A (Below)

Pete Conrad and Alan Bean share a laugh inside the Lunar Module Simulator. (Left)

The Crew in full pressure suits pose inside the Command Module. (Below)

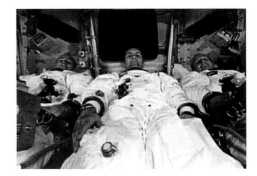

Apollo 12 Lunar Module Intrepid is captured in flight on its way to the Ocean Of Storms.

Pete Conrad descends the ladder to become the third person to walk on the moon.

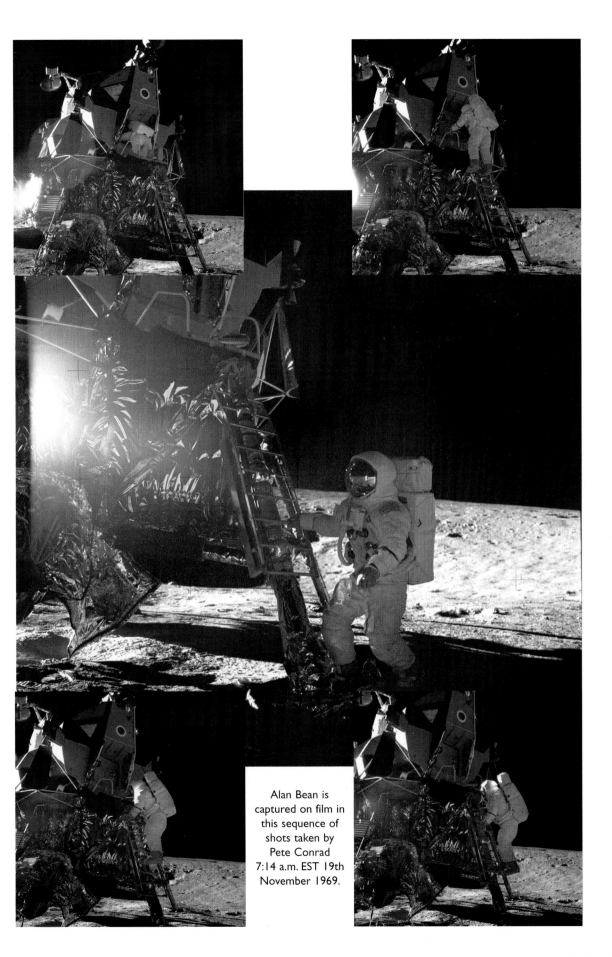

Alan Bean is
captured on film in
this sequence of
shots taken by
Pete Conrad
7:14 a.m. EST 19th
November 1969.

Unpacking the ALSEP from the
back of Intrepid.
The Nuclear Thermoelectric
Generator can be seen in
the center foreground. (Above)

The color television camera from
Apollo 10 was refurbished for the
Lunar surface. The quality of the
pictures were in stark contrast to the
ghostly black and white images
returned by Apollo 11. Sadly the
camera was only functional for a little
over 50 minutes. Controllers in
Houston watch Alan Bean on the big
screen. (Right and Above)

Charles Conrad fulfills a life-long goal. Posing with the flag in The Ocean Of Storms. (Above)

Working at the MESA. The tool carrier can be clearly seen. (Right)

The unfiltered blinding white sunlight illuminates Alan Bean as he carries parts of the ALSEP to its deployment area.

The crew egress from the Command Module after 244 hrs 36 minutes 24 seconds of flight time. The new isolation garments were much easier on them than those worn by Apollo 11's crew. (Above)

Chatting with their families from within the confines of the MQF aboard the USS Hornet. (Right)

Hasselblad Surface Camera

Two calibrated 70mm cameras were used to take a total of approximately 500 black and white and color pictures of the lunar surface prior to, during, and after the EVA's. Color film was used in the first EVA and black and white in the second. Monoscopic and stereoscopic panoramas were taken to record small geologic features. The Apollo Lunar Surface Close-up Camera, especially designed to take close-up stereoscopic color pictures of very small areas on the lunar surface, was used to record selected sites close to the LM.

Optical Properties

Some unusual optical characteristics reported by the astronauts during their EVA's included: (1) green crystals or spots on several rocks, (2) purplish iridescence on at least one rock, (3) the occurrence of several light-gray rocks, and, in one case, a reddish rock, as contrasted with normal occurance of gray to dark gray rocks, and (4) occurrence of light gray fine-grained material at several inches depth (instead of the normal darker gray material) on the rims of at least two craters (Head and Sharp). The astronauts could not detect visually the optical variations in the surface along their traverse (inferred from lunar orbital photographs).

Lunar Surface Materials

As in the Apollo 11 landing, lunar surface material erosion resulted from the Apollo 12 LM descent engine exhaust, and dust was blown from the surface along the flight path. The distance to which the dust was transported will be examined critically from the point of view of possible contamination of the Surveyor III spacecraft. The landing was apparently normal with no excessive stroking of the shock absorbers and without substantial penetration of the footpads into the lunar surface.

After an initial period of adjustment to lunar conditions, the astronauts encountered no unexpected problems in moving about on the surface. Bootprint penetrations were reported to be similar in magnitude to those of the Apollo 11 crew. In the crater containing the Surveyor III spacecraft, the astronauts, burdened with rock samples and a tool carrier, were able to traverse slopes of about 15 degrees. Lunar dust adhered extensively to the astronauts' suits, tools, and other equipment. This was also observed during the Apollo 11 lunar surface activities. The astronauts commented that the Surveyor III was also coated with a thin layer of dust.

Penetration of the lunar surface by various hand tools substantially exceeded that accomplished by the Apollo 11 crew. These included excavation of trenches to a depth of about 20 cm and penetration by a double core tube to a depth of approximately 70 cm. The implication of these results in terms of lunar surface material properties will be closely examined.

In general, the lunar surface material properties are similar to those of the material at the Surveyor spacecraft and Apollo 11 landing sites. The lunar surface material exhibits the properties of a fine-grained, granular material possessing both friction and a slight amount of cohesion. Comparison of photographs taken by the astronauts of the Surveyor III surface sampler tests and footpad imprints with those made by the Surveyor television camera 2½ years ago will give information of considerable interest to lunar material property interpretation.

SYSTEMS PERFORMANCE

Launch vehicle performance was satisfactory throughout its expected lifetime except for the S-IVB slingshot maneuver. The spacecraft systems functioned satisfactorily during the entire mission except for the perturbations caused by an electrical anomaly which occurred shortly after liftoff. Communications were very good except for occasional problems with the High Gain Antenna (HGA). Table 7 summarizes the consumables status at the end of the mission. Tables 8 through 11 summarize the discrepancies encountered during the Apollo 12 Mission.

At 36.5 seconds after liftoff a major electrical anomaly occurred in the space vehicle, attributed to a potential discharge from the clouds through the space vehicle to the ground. The three fuel cells in the SM were disconnected from the main busses, placing main bus A loads on entry battery A and main bus B loads on entry battery B. In addition, numerous alarms and caution and warning lights in the CM cabin were turned on. The signal conditioning equipment dropped out and temperature, RCS propellant secondary quantity, and nuclear particle measurements were permanently lost. A similar incident occurred at 52 seconds GET resulting in tumbling the inertial measurement unit. At T+97.5 seconds the signal conditioning equipment started functioning and by 142.5 seconds GET the crew placed fuel cells 1 and 2 on the line. Fuel cell 3 was placed on the line at 170.5 seconds GET. The electrical power system remained normal throughout the remainder of the mission.

The SM RCS propellant usage exceeded the premission planned rates through most of the mission. This is primarily attributable to two factors: first, a recent procedural change in spacecraft attitude maneuver rate from 0.2 to 0.5 degrees per second; second, higher than planned usage during lunar orbital photography. At all times, the RCS propellant remaining exceeded redline limits by a wide margin.

On several occasions during the mission, communications with the CSM experienced some degradation due to inability of the HGA to hold lock. Two special HGA tests were conducted during the transearth coast to attempt to identify the cause of the anomaly. Results indicate that the problem appears to be associated with the dynamic thermal operation of the antenna, probably in the microwave circuitry in the narrow beam mode.

FLIGHT CREW PERFORMANCE

The Apollo 12 crew performance was outstanding throughout the mission. They exhibited quick thinking and alert reaction during the launch phase emergency. The crew provided a detailed and comprehensive commentary of the lunar surface activities. They were initially cautious in their movements but eventually adopted to the extent that they were able to move (lope) around with great ease. The crew conducted the lunar surface traverse without tiring. Cdr. Conrad fell once but was able to get back up easily from the prone position. He suggested addition of a strap to facilitate a buddy lift. Both crewmen become somewhat thirsty while conducting lunar surface activities in hard suits for an extended period. it was noted that the lunar surface tools were more fragile to handle than anticipated. The crew was able to accomplish all lunar surface requirements.

The crew members took a minimal amount of medication, mostly decongestants to relieve stuffiness which they attributed primarily to lunar dust transferred to the CM by the LM crew's suits and gear. The LMP used sleeping pills prior to two of his rest periods following LM ascent stage separation. Skin cream was used by the Commander to treat a skin rash caused by his biomedical sensors. The crew was exceptionally enthusiastic during all phases of the mission, particularly during the lunar surface extravehicular activities and television transmissions.

All information and data in this report are preliminary and subject to revision by the normal Manned Space Flight Center technical reports.

C.M. Lee

24 NOVEMBER 1969	**APOLLO 12**	**TABLE 1**

1. PRIMARY OBJECTIVES

The following are the OMSF Apollo 12 Primary Objectives.

Perform selenological inspection, survey, and sampling in a mare area
Deploy and activate ALSEP
Develop techniques for a point landing capability
Develop man's capability to work in the lunar environment
Obtain photographs of candidate exploration sites

2. PRINCIPAL AND SECONDARY DETAILED OBJECTIVES

The following are the approved Detailed Objectives:

PRIORITY	DETAILED OBJECTIVES AND EXPERIMENTS
1	Contingency sample collection
2	Lunar surface EVA operations
3	Apollo Lunar Surface Experiments Package (ALSEP 1)
4	Selected sample collection
5	Portable Life Support System Recharge
6	Lunar Field Geology (S-059)
7	Photographs of candidate exploration sites
8	Lunar surface characteristics
9	Lunar environment visibility
10	Landed LM location
11	Selenodetic Reference Point Update
12	Solar Wind Composition (S-080)
13	Lunar multispectral photography (S-158)
14	Surveyor III investigation (SECONDARY OBJECTIVE)
-	Photographic Coverage (SECONDARY OBJECTIVE)
-	Television Coverage (SECONDARY OBJECTIVE)

3. APPROVED EXPERIMENTS

The following are the experiments performed during the mission:

A. S-031   Passive Seismic Experiment (ALSEP 1)
B. S-034   Lunar Surface Magnetometer Experiment (ALSEP 1)
C. S-035   Solar Wind Spectrometer Experiment ((ALSEP 1)
D. S-036   Suprathermal Ion Detector Experiment (ALSEP 1)
E. S.058   Cold Cathode Ion Gauge Experiment (ALSEP 1)
F. M-515   Lunar Dust Detector (ALSEP 1)
G. S-059   Lunar Field Geology
H. S-080   Solar Wind Composition
I. S-158   Lunar Multispectral Photography
J. T-029   Pilot Describing Function

4. SUMMARY

1. It is considered that accomplishment of (1) qualify Apollo 12 as a success. The accomplishment of the Detailed Objectives and Experiments identified in (2) and (3) further enhanced the scientific and technological return of this mission.
2. Other major activities not listed as Detailed Objectives or Experiments:
        * Color TV in CSM
        * Transfer to a non-free return (hybrid) trajectory
        * S-IVB APS evasive maneuver
        * Deorbit of LM ascent stage
3. All detailed objectives were 100% accomplished except:
        * (ALSEP 1) experiments S-036 (SIDE) and S-058 (CCIG): The high voltage is scheduled to be turned on after lunar noon.
        * Television coverage of the lunar surface activities utilizing the color TV on the lunar surface was not accomplished because of a malfunction of the TV camera. This camera is being returned for analysis.

25 NOVEMBER 1969   **APOLLO 12**   **TABLE 2**

ACHIEVEMENTS

* SECOND MANNED LUNAR LANDING MISSION AND RETURN
* SEVENTH SUCCESSFUL SATURN V ON-TIME LAUNCH
* FIRST USE OF THE S-IVB STAGE TO PERFORM AN EVASIVE MANEUVER
* FIRST USE OF A HYBRID TRAJECTORY
* LARGEST PAYLOAD PLACED IN LUNAR ORBIT
* FIRST DEMONSTRATION OF A POINT LANDING CAPABILITY
* FIRST USE OF TWO LUNAR SURFACE EVA PERIODS (ABOUT 4 HOURS EACH)
* FIRST ALSEP DEPLOYED ON THE MOON
* FIRST DEPLOYMENT OF THE ERECTABLE S-BAND ANTENNA
* FIRST RECHARGE OF THE PORTABLE LIFE SUPPORT SYSTEM
* FIRST USE OF GEOLOGISTS TO PLAN A LUNAR SURFACE TRAVERSE IN REAL-TIME
* FIRST DOCUMENTED SAMPLES RETURNED TO EARTH
* FIRST DOUBLE CORE-TUBE SAMPLE TAKEN
* FIRST RETURN OF SAMPLES FROM A PRIOR LUNAR LANDED VEHICLE (SURVEYOR III)
* LONGEST DISTANCE TRAVERSED ON THE LUNAR SURFACE
* LARGEST PAYLOAD EVER RETURNED FROM THE LUNAR SURFACE
* FIRST MULTISPECTRAL PHOTOGRAPHY FROM LUNAR ORBIT
* LONGEST LUNAR MISSION TO DATE

LUNAR MODULE ASCENT STAGE

INERTIAL MEASUREMENT UNIT

WATER TANK

AFT EQUIPMENT BAY

ELECTRONIC EQUIPMENT

GASEOUS OXYGEN TANK (ECS)

HELIUM TANK (ASCENT)

RCS FUEL TANK

RCS QUAD

INGRESS/EGRESS HATCH

RCS OXIDIZER TANK

RCS HELIUM TANK

APS FUEL TANK

25 NOVEMBER 1969   APOLLO 12   TABLE 3

POWERED FLIGHT SEQUENCE OF EVENTS

EVENT	*PLANNED (GET) HR:MIN:SEC	ACTUAL (GET) HR:MIN:SEC
Range Zero 11:22:00.0 EST November 14)	00:00:00:0	00:00:00.0
Liftoff Signal (TB-1)	00:00:00.6	00:00:00.6
Pitch and Roll Start	00:00:12.5	00:00:12.5
Roll Complete	00:00:30.5	00:00:30.5
S-IC Center Engine Cutoff (TB-2)	00:02:15.3	00:02:15.2
Begin Tilt Arrest	00:02:38.5	00:02:41.5
S-IC Outboard Engine Cutoff (TB-3)	00:02:42.5	00:02:41.7
S-IC/S-II Separation	00:02:43.2	00:02:42.4
S-II Ignition (Engine Start Command)	00:02:43.9	00:02:43.1
S-II Second Plane Separation	00:03:13.2	00:03:12.4
Launch Escape Tower Jettison	00:03:18.6	00:03:17.9
S-II Center Engine Cutoff	00:07:41.5	00:07:40.7
S-II Outboard Engine Cutoff (TB-4)	00:09:11.0	00:09:12.3
S-II/S-IVB Separation	00:09:11.9	00:09:13.2
S-IVB Ignition	00:09:13.0	00:09:14.3
S-IVB Cutoff (TB-5)	00:11:29.9	00:11:33.9
Earth Parking Orbit Insertion	00:11:39.9	00:11:43.9
Begin Restart Preparations (TB-6)	02:37:42.0	02:37:43.9
Second S-IVB Ignition	02:47:20.2	02:47:22.6
Second S-IVB Cutoff (TB-7)	02:53:05.0	02:52:03.8
Translunar Injection	02:53:15.0	02:53:13.8

*Prelaunch planned times are based on MSFC Launch Vehicle Operational Trajectory

LUNAR MODULE DESCENT STAGE

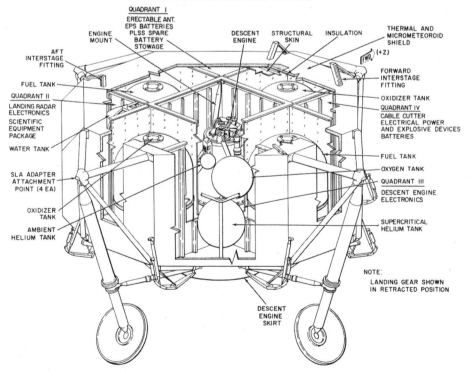

25 NOVEMBER 1969   **APOLLO 12**   **TABLE 4**

MISSION SEQUENCE OF EVENTS

EVENT	*PLANNED (GET) HR:MIN:SEC	ACTUAL (GET) HR:MIN:SEC
Range Zero (11:22:00 EST, November 14)	00:00:00	00:00:00
Earth Parking Orbit Insertion	00:11:40	00:11:44
Second S-IVB Ignition	02:47:20	02:47:23
Translunar Injection	02:53:15	02:53:14
CSM/S-IVB Separation, SLA Panel Jettison	03:12:00	03:18:19
CSM/LM Docking Complete	03:22:00	03-26:58
Spacecraft Ejection from S-IVB	04:07:20	04:13:42
S-IVB APS Evasive Maneuver	04:25:00	**04:27:00
S-IVB Slingshot Maneuver	04:34:25	04:36:19
Midcourse Correction -1	11:53:05	Not Performed
Midcourse Correction -2 (Hybrid Transfer)	30:52:44	30:52:44
Midcourse Correction -3	61:31:14	Not Performed
Midcourse Correction -4	78:31:14	Not Performed
LOI-1 (Lunar Orbit Insertion) Ignition	83:25:18	83:25:23
LOI-2 Ignition	87:44:10	87:48:47
LM Undocking from CSM	107:54:22	107:54:03
CSM Separation Maneuver	108:24:22	108:24:42
LM Descent Orbit Insertion	109:23:00	109:23:39
Powered Descent Initiation	110:19:58	110:20:38
Lunar Landing	110:31:18	110:32: 35
Begin First Extravehicular Activity (EVA-1)	114:40:00	115:10:35
Terminate EVA-1	117:47:00	119:06:38
CSM Plane Change (LOPC-1)	119:47:02	119:47:13
Begin EVA-2	133:15:00	131:32:45
Terminate EVA-2	137:52:00	135:22:00
LM Liftoff	142:08:28	142:03:47
Coelliptic Sequence Initiation Maneuver	142:58:05	143:01:51
Constant Differential Height Maneuver	143:56:28	144:00:02
Terminal Phase Initiation Maneuver	144:36:25	144:36:29
LM/CSM Docking	145:40:00	145:36:22
LM Jettison	147:57:00	147:59:30
CSM Separation Maneuver	148:02:09	148:04:30
Ascent Stage Deorbit Maneuver	149:24:41	149:28:14
Ascent Stage Lunar Impact	149:52:51	149:55:17
CSM Plane Change (LOPC-2)	159:01:46	159:04:45
Transearth Injection Ignition	172:21:15	172:27:16
Midcourse Correction -5	187:23:24	188:27:14
Midcourse Correction -6	222:21:48	Not Performed
Midcourse Correction -7	241:21:48	241:21:57
CM/SM Separation	244:06:48	244:07:21
Entry Interface (400,000 feet)	244:21:48	244:22:19
Landing	244:35:23	244:36:24

*Prelaunch planned times are based on MSFC Launch Vehicle Operational Trajectory and MSC Spacecraft Operational Trajectory

**This time is approximate. MSFC does not have actual times.

24 NOVEMBER, 1969   APOLLO 12   TABLE 5

TRANSLUNAR AND TRANSEARTH MANEUVER SUMMARY

MANEUVER	GROUND ELAPSED TIME (GET) AT IGNITION (Hr:Min:Sec) PRE-LAUNCH	REAL-TIME	ACTUAL	BURN TIME (seconds) PRE-LAUNCH PLAN	REAL-TIME PLAN	ACTUAL	VELOCITY CHANGE (feet per second - fps) PRE-LAUNCH PLAN	REAL TIME PLAN	ACTUAL	GET OF CLOSEST APPROACH HT (NM) CLOSEST APPROACH PRE-LAUNCH PLAN	REAL-TIME PLAN	ACTUAL
TLI (S-IVB)	02:47:19.8	02:47:20.6	02:47:20.6	345	344	344	10,510	10,515	10,515	83:25:51 1851	83:43:15 457	83:43:15 457
MCC-1	11:53:04.8	—	N.P.	0	—	N.P.	0	—	N.P.	83:25:51 1851	—	N.P. N.P.
MCC-2 (SPS)	30:52:43.7	30:52:43.7	30:52:43.7	10	8.8	8.8	68.8	61.7	61.8	82:28:59 59	83:28:30 60	83:28:26 60
MCC-3	61:31:13.6	—	N.P.	0	—	N.P.	0	—	N.P.	83:28:59 59	—	N.P. N.P.
MCC-4	78:31:13.6	78:25:18.2	N.P.	0	16.7	N.P.	0	1.0	N.P.	83:28:59 59	83:28:27 60	N.P. N.P.

MANEUVER	PRE-LAUNCH	REAL-TIME	ACTUAL	PRE-LAUNCH PLAN	REAL-TIME PLAN	ACTUAL	PRE-LAUNCH PLAN	REAL TIME PLAN	ACTUAL	GET ENTRY INTERFACE (EI) VELOCITY (fps) AT EI FLIGHT PATH ANGLE AT EI PRE-LAUNCH PLAN	REAL-TIME PLAN	ACTUAL
LUNAR ORBIT	Lunar orbit maneuvers are summarized on a separate table.											
TEI (SPS)	172:21:14.7	172:27:16.1	172:27:16.1	129	130	130	3,036	3042.3	3042.4	244:21:48 36,116 -6.5	244:21:27 36,116 -6.5	244:21:14 36,116 -6.69
MCC-5	187:23:25.6	188:27:13.7	188:27:13.7	0	4.53	4.54	0	2.0	2.0	244:21:48 36,116 -6.5	244:22:06 36,116 -6.50	244:22:01 36,116 -6.49
MCC-6	222:21:47.5	—	N.P.	0 -	—	N.P.	0	—	N.P.	244:21:48 36,116 -6.5	—	244:22:32 36,116 -6.01
MCC-7	241:21:47.5	241:21:57.4	241:21:57.4	0	5.23	4.69	0	2.4	2.1	244:21:48 36,116 -6.5	244:22:18 36,116 -6.49	244:22:19 36,116 -6.47

N.P. - Not Performed

APOLLO CM INTERIOR CONFIGURATION
LHEB, LHFEB, UEB, & LEFT AB

-Z   -Y
X

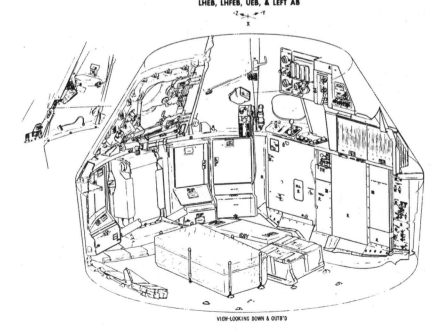

VIEW-LOOKING DOWN & OUTB'D

24 NOVEMBER, 1969   APOLLO 12   TABLE 6

LUNAR ORBIT MANEUVER SUMMARY

MANEUVER	GROUND ELAPSED TIME (GET) AT IGNITION (Hr:Min:Sec)			BURN TIME (seconds)			VELOCITY CHANGE (feet per second - fps)			APOLUNE/PERILUNE RESULTANT (NAUTICAL MILES)		
	PRE-LAUNCH PLAN	REAL-TIME PLAN	ACTUAL	PRE-LAUNCH PLAN	REAL-TIME PLAN	ACTUAL	PRE-LAUNCH PLAN	REAL TIME PLAN	ACTUAL	PRE-LAUNCH PLAN	REAL-TIME PLAN	ACTUAL
Lunar Orb Insertion LOI-1	83:25:18.2	83:25:22.7	83:25:22.7	355.4	358	352	2889.5	2889.9	2889.3	170.2 / 59.9	169.3 / 62.3	168.8 / 62.6
Lunar Orb Insertion (LOI-2)	87:44:10.0	87:48:47.4	87:48:47.4	17.6	17.1	17.0	169.6	165.5	165.2	66.0 / 54.2	66.2 / 54.1	66.1 / 54.3
CSM/LM Sep CSM Result	108:24:22.0	108:24:42	108:24:42	15.8	15.96	15.43	2.5	2.5	2.4	64.2 / 55.7	63.7 / 56.4	63.5 / 56.3
Descent Orbit Insertion	109:23:00.0	109:23:39.4	109:23:39.4	28.2	28.96	28.97	72.0	72.4	72.4	59.3 / 8.3	60.5 / 8.3	60.6 / 8.1
Powered Descent Initiation	110:19:58.0	110:20:37.4	110:20:37	680	680.53	715	6611	6618.7	—	0 / 0	0 / 0	0 / 0
CSM Plane Change (LOPC-1)	119:47:02.0	119:47:12.5	119:47:12.5	19.4	18.14	18.2	372.4	349.7	350.0	62.7 / 56.9	62.5 / 57.3	62.5 / 57.5
Ascent To Insertion	142:08:27.9	142:03:47	142:03:47	430.0	452.6	—	6046.2	6059.5	6057.4	45.3 / 9.0	46.3 / 8.8	46.3 / 8.8
Coelliptic Sequence Initiate	142:58:05.2	143:01:51	143:01:50.6	45.3	—	41.1	50.3	49.0	45.2	45.6 / 44.6	— / —	51.0 / 41.5
Constant Delta	143:56:27.5	144:00:01.5	144:00:01.5	0	12.66	12.5	0	14.0	13.8	— / —	44.2 / 40.4	44.4 / 40.4
Terminal Phase Initiate	144:36:25.7	144:34:—	144:36:29.4	22.1	—	25.75	24.6	25.0	28.5	61.9 / 44.2	— / —	60.2 / 43.8
Final Braking (Docking)	145:21:57.1	145:20:59	145:19:29.3	4.2	—	—	4.7	32.	36.3	59.8 / 59.5	— / —	— / —
CSM//LM Sep CSM Result	148:02:08.6	—	148:04:30	3.1	—	—	1.0	—	—	60.0 / 59.8	— / —	— / —
LM Ascent Stage Deorbit	149:24:41.2	149:28:14	149:28:14	83.8	81.2	83.45	189.7	190.9	196.3	59.9 / -64.4	58 / -62.5	57.7 / -66.3
CSM Plane Change	159:01:46.0	159:04:44.8	159:04:44.8	18	19.1	19.2	360	381.3	382	61.2 / 57.8	64 / 56.5	64.1 / 56.7

APOLLO COMMAND MODULE INTERIOR

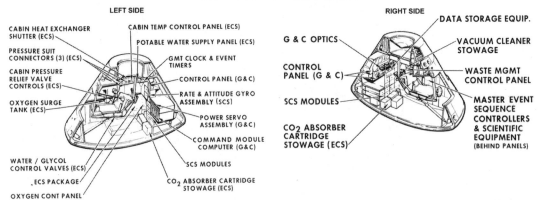

LEFT SIDE

CABIN HEAT EXCHANGER SHUTTER (ECS)
PRESSURE SUIT CONNECTORS (3) (ECS)
CABIN PRESSURE RELIEF VALVE CONTROLS (ECS)
OXYGEN SURGE TANK (ECS)
WATER / GLYCOL CONTROL VALVES (ECS)
ECS PACKAGE
OXYGEN CONT PANEL

CABIN TEMP CONTROL PANEL (ECS)
POTABLE WATER SUPPLY PANEL (ECS)
GMT CLOCK & EVENT TIMERS
CONTROL PANEL (G & C)
RATE & ATTITUDE GYRO ASSEMBLY (SCS)
POWER SERVO ASSEMBLY (G & C)
COMMAND MODULE COMPUTER (G & C)
SCS MODULES
CO_2 ABSORBER CARTRIDGE STOWAGE (ECS)

RIGHT SIDE

G & C OPTICS
CONTROL PANEL (G & C)
SCS MODULES
CO_2 ABSORBER CARTRIDGE STOWAGE (ECS)

DATA STORAGE EQUIP.
VACUUM CLEANER STOWAGE
WASTE MGMT CONTROL PANEL
MASTER EVENT SEQUENCE CONTROLLERS & SCIENTIFIC EQUIPMENT (BEHIND PANELS)

24 NOVEMBER 1969   APOLLO 12   TABLE 7

CONSUMABLES SUMMARY AT END OF MISSION

CONSUMABLE		LAUNCH LOAD	PRELAUNCH PLANNED REMAINING	ACTUAL REMAINING
CM RCS PROP (POUNDS/PERCENT)	D	208/100	194/93.3	Not Available
SM RCS PROP (POUNDS/PERCENT)	U	1,233/100	540/43.8	272/20.6
SPS PROP (POUNDS/PERCENT)	TK	40,614/100	3339/8.22	3473/8.55
SM HYDROGEN (POUNDS/PERCENT)	U	54.3/100	8.0/14.8	9.7/17.3
SM OXYGEN (POUNDS/PERCENT)	U	634/100	164/25.9	150/23.2
LM RCS PROP (POUNDS/PERCENT)	U	549/100	*173/30.2	*110/20.1
LM DPS PROP (POUNDS/PERCENT)	U	18,226/100	**1491/8.18	**705/3.87
LM APS PROP (POUNDS/PERCENT)	U	5,182/100	*320/6.18	*285/5.50
LM A/S OXYGEN (POUNDS/PERCENT)	T	4.72/100	*3.28/69.5	*4.01/85.0
LM D/S OXYGEN (POUNDS/PERCENT)	T	48.9/100	**17.0/34.8	**23.7/48.5
LM A/S WATER (POUNDS/PERCENT	T	85.0/100	*45.0/53.0	*45.0/53.0
LM D/S WATER (POUNDS/PERCENT)	T	252/100	**88.0/34.9	**80.8/32.1
LM A/S BATTERIES (AMP-HRS/PERCENT)	T	592/100	*165/27.9	*185/31.2
LM D/S BATTERIES (AMP-HRS/PERCENT)	T	1,600/100	**452/28.2	**576/36.0

D- Deliverable Quantity   * At LM Ascent Stage Lunar Impact
U- Usable Quantity   ** At LM Ascent Stage Liftoff
TK- Tank Quantity
T- Total Quantity

24 NOVEMBER 1969   APOLLO 12   TABLE 8

SA-507 LAUNCH VEHICLE

DISCREPANCY SUMMARY

1.   THE S-IVB O2H2 BURNER OXIDIZER SHUTOFF VALVE FAILED TO CLOSE WHEN THE IU ISSUED A CLOSED SIGNAL AT 9958.8 SECONDS, BUT WAS CLOSED BY GROUND COMMAND.

2.   LOW FREQUENCY OSCILLATIONS (16 HZ) WERE EXPERIENCED DURING S-II POWERED FLIGHT BEGINNING AT 340 SECONDS AND OCCURRING AT 3 OTHER DISTINCT PERIODS DURING THE BURN.

3.   A CCS DOWNLINK DROPOUT WAS EXPERIENCED ON OMNI ANTENNA AT 19,255 SECONDS. PERFORMANCE WAS NOMINAL ON LOW GAIN AND HIGH GAIN ANTENNA. UPLINK LOCK WAS LOST AND NEVER REGAINED AFTER 26,786 SECONDS.

4.   AN ELECTRICAL PHENOMENON OCCURRED AT APPROXIMATELY 36 SECONDS THAT WAS INDICATED BY A TRANSIENT RESPONSE IN APPROXIMATELY 60 LAUNCH VEHICLE MEASUREMENTS. A RELATIONSHIP HAS NOT BEEN ESTABLISHED BUT OTHER OCCURRENCES INCLUDED

    a. AT APPROXIMATELY 35.53 SECONDS THE IU LVDA PITCH GIMBAL CROSSOVER COUNTERS OUTPUT INDICATED A CHANGE IN EXCESS OF THE ACCEPTABLE LIMIT OF 0.4 DEG. THE COMPUTER RESPONDED PROPERLY BY REUSING THE LAST COUNTER READING.

    b. AT APPROXIMATELY 36.17 SECONDS BIT 1 IN MODE CODE 24 WAS SET BECAUSE OF THE REDUNDANT ACCELEROMETER COUNTERS DIFFERED BY APPROXIMATELY 9 COUNTS. THE COMPUTER USED THE MORE REASONABLE VALUE.

5.   TRACKING VECTORS INDICATE THE S-IVB/IU WAS PLACED IN A VERY ELLIPTICAL EARTH ORBIT RATHER THAN THE PLANNED HELIOCENTRIC ORBIT.

24 NOVEMBER 1969   APOLLO 12   TABLE 9

COMMAND/SERVICE MODULE-108 DISCREPANCY SUMMARY

1. ALL ELEMENTS ON THE DSKY (DISPLAY KEYBOARD) LIT UP INTERMITTENTLY (PRELAUNCH).

2. TUNING FORK INDICATION INTERMITTENT ON MISSION TIMERS. CENTRAL TIMING EQUIPMENT VERIFIED OPERATIONAL (PRELAUNCH).

3. LEAK IN CRYOGENIC HYDROGEN TANK LINE INTERFACE TO INNER TANK (PRELAUNCH).

*4. AT 36.5 SECONDS GET — FUEL CELLS DROPPED OFF BUSES, LOSS OF AC BUSES AND EVENT TIMER, TRANSIENT ON LAUNCH VEHICLE BUS DUE TO POTENTIAL DISCHARGE TO GROUND. FOUR SKIN TEMPERATURES ON SERVICE MODULE FAILED.

5. THE FOUR SERVICE MODULE REACTION CONTROL SYSTEM BACKUP QUANTITY MEASUREMENT PRESSURE/TEMPERATURE DEVICES FAILED.

*6. AT 52 SECONDS GET — INERTIAL MEASUREMENT UNIT LOSS, OTHER PULSE CODE MODULATION INDICATION AND LAUNCH VEHICLE TRANSIENTS INDICATED ANOTHER DISCHARGE POTENTIAL.

7. HELIUM QUAD B AND QUAD A SECONDARY PROPELLANT TALK BACKS INDICATED BARBERPOLE AT S-IVB SEPARATION.

8. AFTER INSERTION STABILIZATION AND CONTROL SYSTEM LOGIC 3-4 MAIN A WAS FOUND TO BE OPEN.

9. AT 31:40 GET — MASTER ALARM WITH NO ANNUNCIATOR.

10. CARBON DIOXIDE PARTIAL PRESSURE INDICATOR READ LOW LEVEL.

*11. UP AND DOWNLINK COMMUNICATIONS SIGNAL STRENGTHS BELOW EXPECTED LEVELS FOR HIGH GAIN ANTENNA OPERATION IN NARROW BEAM AUTO TRACK DURING SEVERAL PERIODS IN LUNAR ORBIT.

12. CALCULATED ENVIRONMENTAL CONTROL SYSTEM AND FUEL CELL OXYGEN USAGE DIFFERED FROM TELEMETERED QUANTITY.

13. HIGH OXYGEN FLOW WITHOUT MASTER ALARM SEVERAL TIMES DURING FLIGHT.

14. LOOSE PIECE OF MATERIAL OBSERVED BY LM CREW AT FINAL DOCKING.

*15. OPTICS COUPLING DISPLAY UNIT INDICATED OPTICS MOVEMENT WHEN OPTICS IN "ZERO OPTICS" MODE — OCCURRED SEVERAL TIMES.

16. QUAD D HELIUM MANIFOLD PRESSURE TRANSDUCER READ IMPROPERLY.

*17. CLOGGED URINE DUMP FILTERS,

18. RADIATOR HEATERS CAME ON AT -7° INSTEAD OF -15°F.

* MOST SIGNIFICANT FLIGHT PROBLEMS

LUNAR MODULE ASCENT STAGE INTERIOR VIEW LOOKING FORWARD

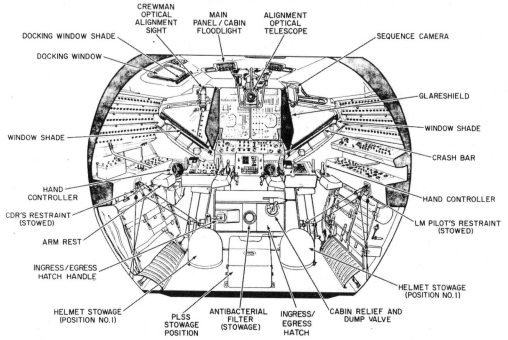

24 NOVEMBER 1969   APOLLO 12   TABLE 10

LUNAR MODULE-6 DISCREPANCY SUMMARY

1. FLOODLIGHT DID NOT TURN OFF WHEN LM TUNNEL HATCH WAS CLOSED BUT SWITCH TURNED FLOODLIGHTS OFF WHEN SWITCH OPERATED BY HAND.

2. LM DOOR CLOSED WHILE COMMANDER WAS ON LUNAR SURFACE AND LUNAR MODULE PILOT WAS IN LM,

*3. WATER ACCUMULATED IN COMMANDER'S SUIT LOOP AND BOOTS.

4. MASTER ALARM DURING ASCENT AND IMMEDIATELY AFTER CUTOFF,

*5. LM TRACKING LIGHT NOT WORKING AT 144:26 GET.

6. CARBON DIOXIDE PARTIAL PRESSURE CAUTION AND WARNING WHILE BEHIND MOON ON REVOLUTION 31.

7. REACTION CONTROL SYSTEM MAIN SHUTOFF VALVE, SYSTEM A INDICATOR SHOWED CLOSED DURING INITIAL ACTIVATION (FINAL PHASES OF APS BURN). PROPER AFTER RECYCLE.

*8. POOR VHF (VERY HIGH FREQUENCY) RELAY TO CSM DURING LM ASCENT.

9. COMMUNICATION TONES (KEYING) DURING EXTRAVEHICULAR ACTIVITY.

* MOST SIGNIFICANT FLIGHT PROBLEMS

24 NOVEMBER 1969   APOLLO 12   TABLE 11

CREW/EXPERIMENT EQUIPMENT DISCREPANCY SUMMARY

*1. TELEVISION CAMERA INOPERATIVE AFTER REMOVAL FROM LM

*2. 16mm SEQUENCE CAMERA STOPPED OPERATING DURING LM ASCENT

3. EQUIPMENT TRANSFER BAG DIFFICULTY WITH LUNAR EQUIPMENT CONVEYOR INTERFACE DURING INITIAL PHASE OF SECOND EXTRAVEHICULAR ACTIVITY PERIOD

4. RADIOISOTOPE THERMOELECTRIC GENERATOR FUEL ELEMENT DIFFICULT TO REMOVE FROM CASK

5. PASSIVE SEISMIC EXPERIMENT DEPLOYMENT DIFFICULT

6. SUPRATHERMAL ION DETECTOR WOULD NOT REMAIN UPRIGHT

7. SHORTING PLUG AMMETER DID NOT INDICATE CURRENT PRIOR TO ACTIVATION OF SHORTING PLUG SWITCH

*8. TEFLON BAGS CRACKED AT FOLD WHEN FOLDED WHILE ON LUNAR SURFACE

9. RETAINING CLIPS FOR SAMPLE BAG WOULD NOT HOLD BAG

*10. DURING LANDMARK TRACKING 500mm CAMERA COUNTER DID NOT AGREE WITH CREW COUNT. ALSO, BACK OF MAGAZINE CAME OFF

* MOST SIGNIFICANT FLIGHT PROBLEMS

CONFIDENTIAL

NATIONAL AERONAUTICS AND SPACE ADMINISTRATION

APOLLO 12
TECHNICAL
CREW DEBRIEFING
(U)

DECEMBER 1, 1969

PREPARED BY:
MISSION OPERATIONS BRANCH
FLIGHT CREW SUPPORT DIVISION

```
GROUP 4
Downgraded at 3-year
intervals; declassified
after 12 years
```

CLASSIFIED DOCUMENT - TITLE UNCLASSIFIED

This material contains information affecting the national defense of the United States within the meaning of the espionage laws, Title 18, U. S. C., Secs. 793 and 794, the transmission or revelation of which in any manner to an unauthorized person is prohibited by law.

MANNED SPACECRAFT CENTER
HOUSTON, TEXAS

CONFIDENTIAL

SECURITY CLASSIFICATION

The material contained herein has been transcribed into a working paper in order to facilitate review by interested MSC elements. This document, or portions thereof, may be declassified subject to the following guidelines:

Portions of this document will be classified CONFIDENTIAL, Group 4, to the extent that they: (1) define quantitative performance characteristics of the Apollo Spacecraft, (2) detail critical performance characteristics of Apollo crew systems and equipment, (3) provide technical details of significant launch vehicle malfunctions in actual flight or reveal actual launch trajectory data, (4) reveal medical data on flight crew members which can be considered privileged data, or (5) reveal other data which can be individually determined to require classification under the authority of the Apollo Program Security Classification Guide, SCG-11, Rev. 1, 111166.

1.0 SUITING AND INGRESS

CONRAD — There were no noted differences from the published procedure.

GORDON — I agree. It was all nominal.

CONRAD — It was like a CDDT.

2.0 STATUS CHECKS AND COUNTDOWN

CONRAD — There were no noted differences from published procedures.

BEAN — The S-band antenna was in the wrong position in the spacecraft and we changed that before lift-off. The other is the water under the BPC.

2.5 Distinctions of Sounds in the Launch Vehicle Sequence, Countdown to Lift-Off

CONRAD — We noted no sounds whatsoever from the boosters. It was extremely quiet up there. The only noteworthy condition was discussed the previous evening with George Page in the control room. We had discussed the heavy rain and the fact they were going to roll back the White Room. I was concerned about water and he rested assured that the BPC was waterproof and that it was perfectly safe. My other concern was that the upward firing SM RCS thrusters were all going to be full of water. Everybody concluded that this was no problem. During the countdown with the wind blowing up there, it was obvious to me that water was leaking between the BPC and the spacecraft. I could see water on my two windows - windows 1 and 2. We experienced varying amounts passing across these windows, dependent on how heavily it was raining. These were the only things noted up to lift-off.

3.0 POWERED FLIGHT

CONRAD — The three of us felt that the noise level was lower than anticipated. None of us wore anything except the normal helmets. We had no earpieces or ear-tubes as some previous crews had used. Throughout the atmospheric portion of powered flight where you can expect high noise levels, we didn't find it particularly noisy. The flight was extremely normal for the first 36 seconds and after that it got very interesting.

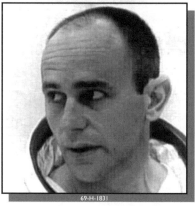

3.1 S-IC IGNITION

GORDON — I want to say there were very good physical cues and there was no doubt in anybody's mind that lift-off had occurred. The cues were there, and you knew when lift-off had occurred regardless of the clock and the lift-off call from LCC.

3.2 LIFT-OFF

CONRAD — I'll run quickly through the launch program here.

3.3 Launch Vehicle Lights

CONRAD — The launch vehicle lights went out as advertised in the proper sequence. They are slightly staggered and it was about a second and a half before lift-off - all the lights were out and we had lift-off.

3.4 Roll Program

CONRAD — The roll program was on time.

3.5 Pitch Program

CONRAD — The pitch program was on time.

69-H-1459

3.6 Rate Changes

CONRAD — I didn't notice the rate changes, because, at 36 seconds, I first noticed that something had happened outside the space craft. I was aware of a white light. I knew that we were in the clouds; and, although I was watching the gauges I was aware of a white light. The next thing I noted was that I heard the MASTER ALARM ringing in my ears and I glanced over to the caution and warning panel and it was a sight to behold. There's a little disagreement among us and I'll have to look at the tapes, but my recollection of what I called out was three FUEL CELL lights, both AC 1 BUS and AC 1 OVERLOAD, FUEL CELL DISCONNECT, MAIN A and B BUS OVERLOAD lights, and I was not aware of AC 2 lights. Dick thought they were on. I don't think they were because I remember thinking that the only lights that weren't on of the electrical system was AC 2 and maybe I ought to configure for an AC BUS 1 out.

69-H-1469

GORDON — Let me make a comment here. A considerable length of time elapsed between the time those lights came on and when Pete read them off to the ground. I can't swear positively that they were all on. To help Al, my usual habit was, when any light came on during the boost phase, to read it out so that he didn't have to be concerned with which light it was. My recollection is that when I first glanced up there I didn't read any of them to him, but I scanned all of them and the only thing I said to him was, "Al, all the lights are on." I am under the impression that at one time, or initially at least, they were all on. Pete read it out quite a bit later. We'd talked about it and you read them out a minute or so later.

69-H-1470

CONRAD — I went back to my gauges and ascertained that everything was running on my side. About this time, the platform tumbled. The next thing I noted was that the ISS light was on. This was obvious because my number 1 ball was doing 360's. I even took the time to peek under the card and we had a PROGRAM ALARM light, the GIMBAL light, and a NO ATT light. So we lit up just about everything in the spacecraft. Since that time, we found out that we were hit a second time and that's probably what did the platform in. When the platform went, Al was telling me that we had voltage on all buses and that we had all buses. I remember him telling me that the voltage was low - 24 volts.

BEAN — We got all the lights. I didn't have an idea in the world what happened. My first thought was that we might have aborted, but I didn't feel any g's, so I didn't think that was what had happened. My second thought was that somehow the electrical connection between the command module and the service module had separated, because all three fuel cells had plopped off and everything else had gone. I immediately started working the problem from the low end of the pole. I looked at both ac buses and they looked okay, so that was a little confusing.

GORDON — You looked at the voltage meters, not the lights.

S69-58879

S69-58884

69-H-1862

BEAN — Yes, that's right. I looked at the volts - the lights were on. I looked at the voltage and the voltage on all phases was good. This was a little confusing. Usually when you see an ac over voltage light, either a inverter goes off or you have one of the ac phases reading zero and you have to take the inverter off. In this case, they all looked good and that was a bit confusing. I switched over and took a look at the main buses. There was power on both, although the voltage was down to about 24, which was a lot lower than normal. I looked at the fuel cells and they weren't putting out a thing. I looked at the battery buses and they were putting out the same 24 volts. They were hooked into the mains and it turned out that they were supplying the load. As I did this, I kept telling Pete we had power on all these buses. One of the rules of space flight is you don't make any switch-a-roos with that electrical system unless you've got a good idea why you're doing it. If you don't have power at all, you might change a couple of switches to see what will happen. When you have power and everything is working, you don't want to switch too much. I didn't have any idea what had happened. I wasn't aware anything had taken place outside of the spacecraft. I was visualizing something down in the electrical systems.

CONRAD — We had a crew rule to handle electrical emergencies. Al did not do any switching without first telling me what he was going to do. When he told me that we had power on all buses, I remember making the comment to him not to do anything until we got through staging.

BEAN — Yes, that is right. I didn't have any ideas anyway. I knew we had power, so I didn't want to make any changes. I figured we could fly into orbit just like that and that's exactly what we did. The ground came up a little later and said to put the fuel cells back on the line. I was a little hesitant about doing that, because I didn't understand that we had been hit by lightning. I gave it a go and, sure enough, things started working very well after that.

CONRAD — Because I could see outside, I made the comment to them several times. I told the ground that I thought we had been hit by lightning. I was the only one that had any outside indications. Dick didn't note anything over his little hole in his center window. I was the only one who noticed anything and that was only the first time. I was aware that something external to the spacecraft had happened. I had the decided impression that I not only saw it, but felt it and heard it.

69-H-1858

BEAN — I think the one thing I should have done was put battery C on both buses. I don't think you're going to hurt yourself doing this. I would not have tried to reset the fuel cells or anything else any faster than when the ground called up, because everything was working fairly well and we were in a critical flight phase. I didn't want to take a chance of taking out a bus with a bad switch.

GORDON — I think it was a smart decision. We've all learned that by arbitrarily switching the electrical system around, you can get yourself into more trouble.

BEAN — You could lose the whole ballgame and we had the whole ballgame. We were in pretty good shape.

69-H-1857

CONRAD — I never considered any kind of an abort. The only concern that passed my mind was winding up in orbit with that dead spacecraft. As far as I could see, as long as Al said he had power on the buses and the COMM was good, we'd press on. We had a long time to go before we could do a mode II and so the thought never crossed my mind about aborting at any point. I wanted to make sure we had enough time to psyche it out. The main concern I had was getting through staging where we got the g levels back down again and we had a little more time to sort out what was going on. So we can back up and say that we heard cabin pressure venting. Engine gimbaling was apparent to everybody in the first 15 or 20 seconds of powered flight, especially going by the tower. The "Tower Clear" call was loud and clear.

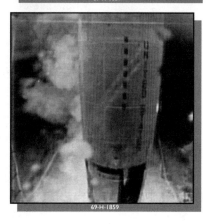

69-H-1859

GORDON — Let me inject something here. During this time, we didn't have any roll or any calls from the ground or anything. I didn't hear 1-B called out; during all the confusion of all the lights, I did not throw the RCS propellant command to RCS. I missed that switch. The next thing I recall being called from the ground after this electrical problem got sorted out in everybody's mind was a 1-C call and I thought, "I've got to get over here," and I turned off the RCS propellant command at that time. Then, in 2 minutes, I got the EDS. I guess the rules say that when you lose a fuel cell you turn off the EDS; but there was so much confusion at that time that I just got the EDS functions at the 2 minute time.

CONRAD — I guess the most serious thing was the second lightning strike, which we weren't aware of. I was under the impression we lost the platform simply because of low voltage but apparently that's not the case. Apparently we got hit a second time at that point. Dick's right with that EDS AUTO enabled. All we needed to do was blow a battery off the line and I have a decided impression we would have gotten an AUTO abort.

69-H-1861

GORDON — No, you need two of them.

CONRAD — I think because of previous crew briefings, there were no surprises in S-IC staging. We got all the good things that most people mentioned and so much for that.

3.10 S-II Engine Ignition

CONRAD — S-II engine ignition was smooth and the ride throughout the S-II flight was as advertised except for a 1 to 2 Hz vibration we felt throughout the whole burn.

GORDON — It was not necessarily longitudinal and was just a general vibration.

CONRAD — I don't really know what it was. I was just aware of it. I don't think any of us noticed the mixture ratio shift on the S-II. We did notice it on the S-IVB. The tower and BPC went as advertised; but, when they did they unloaded a whole pile of water on the spacecraft again and

S69-60068

this water streaked down the windows and froze immediately. At the same time, the water picked up particles from the LET jettison motor and deposited a white ash in the form of oil droplets and streaks all over windows 1, 2, and 3; a little on 4, and none on 5, which was our best window. The ice sublimated later en route to the Moon after TLI, but it left white deposits in the form of spiderweb-like things in the corner crevices and as a white deposit on the windows. The S-II center engine shutdown came as advertised and S-II shutdown came on time.

3.15 S-IVB IGNITION

CONRAD — S-II and S-IVB separations were smooth and normal. S-IVB ignition was smooth. The ride on the S-IVB was very nice. Communications throughout the whole launch were outstanding. One other anomaly during all this mess with the lights. We had several MASTER ALARMS after the initial one that turned out to be the CO_2 sensor. It flashed on and off several times; when we looked at the gauge, it was just cycling from zero to full scale. It finally quivered one more time and gave up the ghost and died somewhere in the powered flight, never to be heard from again for the rest of the mission. The remaining controls displays operated as normal until we were in flight, when we noticed we had a failure of our service module RCS quantity gauge. This must have been a gauge failure because it failed for all four quads. As far as any other sensations through powered flight were concerned, we felt that they were all normal per previous crew briefings.

GORDON — Number 1 window picked up some ice right in the center sometime after launch. The ice looked like it was about 3 inches in diameter and was located on the inside of the outer pane. It actually sublimated off during our first night and was gone the morning of the second day.

CONRAD — I don't know where that came from. I don't have any idea and there wasn't anything like it on the other windows.

<table>
<tr><td>

4.0 EARTH ORBIT AND SYSTEMS CHECKOUT

</td><td>

CONRAD — There was no evaluation of the insertion parameters. We didn't have a computer or we didn't have a platform. So, we started our normal post-insertion systems configuration checks. We knew we had to do a P51 immediately upon hitting darkness, so Dick started hustling down the LEB during daylight. The ground was going through a full-scale review so it could be determined what happened and whether they were going to give us the GO for TLI or not. We had to get the platform up. I knew they wanted to look at that and I was sure they were going to come up with several other things for us to look at; which they did. All of them we agreed with; all of them were no strain to pick out. I think the ground organized it very well by passing us the pages of the checklist they wanted to run over. They wanted a quick P52. We knew they were going to ask for a second P52 to get a look at the platform. They wanted an MTV check of VERB 74 on the erasable and I presumed they looked at many other systems on the telemetry. We got the ORDEAL up and operating. Dick jettisoned the optics covers.

</td></tr>
</table>

GORDON — No debris was noticed as I saw the optics covers go through the sextant. I heard them go. I just had the optics control full-right so that the shaft angle was increasing. By the time I got over to look through the telescope, the cover had already gone. I did not see it go, but I was positive that both covers, in fact, had left the spacecraft.

CONRAD — The S-IV responded as it should, there was no doubt in my mind without a COAS in the window that we were in local horizontal torquing. The unstorage went per checklist. The COMM was good and per previous simulations. Except for the additional systems checks we went through, the TLI PREP went exactly according to checklist. We'll each cover the one anomaly we noted in the CO_2 sensor and the two anomalies in the RCS gauge somewhere along the line. I was very surprised I couldn't hear the RCS thrusters fire when we did the pulse check. The ground came back and assured me that Neil couldn't hear them either. The reason I pulsed several times on it was that I was still worried about the water in those thrusters. I wasn't convinced, in my mind, that we had not frozen some thrusters full of ice as there was water on the windows. Everybody thought it would disappear and it hadn't. I was concerned about those service module ECS thrusters, but the ground assured me they were working okay and it was all right with us. We all had a fullness in the head and, other than that, I don't think anyone suffered any nausea, vertigo, or any other symptom.

CONRAD — Unless someone wants to comment on that, I'll let Dick talk about the P51.

BEAN — I think we were all pretty careful and I had the feeling that if I had moved around a lot, I could have gotten dizzy. But I never did. Everyone was pretty careful and after about a day, it didn't make any difference. We were doing anything we wanted. I don't know if anyone mentioned it before, but your head shape changes. I looked over at Dick and Pete about 2 hours after insertion and their heads looked as if they had gained about 20 pounds.

BEAN — Did you mention that inside some of those windows, not only did we have that icy stuff, but we have that fluffy cotton-like material?

CONRAD — That cotton-like material was on the outside and that's what I called a spider web. It was actually ice to begin with, and when the ice sublimated, it just left those filmy deposits.

GORDON — I'm sure it was the deposits we picked up during BPC jettison. I don't know where it was from, but it was there.

BEAN — You did a gimbal motor check too, didn't you?

GORDON — Yes, MTV check. I guess Pete has covered all the extra things we did as far as the systems are concerned: E-memory dump, self-check, TVC check, and getting ready for TLI. Before I get in the P51, I might comment on the nausea, vertigo, and other disturbances I fully expected in the head, which was experienced in other flights by other people. I guess my comment is, that it was there and was recognized and there was no concern. After the first night, it was gone and never returned. I was hustling around the LEB area quite a bit because I knew we had a lot to do before we could get a commitment for TLI. Even though I was not as careful in the manner in which I moved, I was not concerned about moving out ahead or underneath the couches and all that. I did this without any sensations of vertigo or nausea whatsoever.

I think the more you think on these things, the more susceptible you are to it. So I just forgot about them. Knowing I would have to do a P51, I grabbed the star charts out of the storage area, along with the rest of the unstored items that had to be accomplished and started looking on the star chart as to where we might pick up some stars and I just sat there waiting for darkness. I heard this comment before and was fully aware of it. When we got to darkness, I looked in the telescope and didn't see anything. I was concerned at that time that I hadn't gotten rid of the optics covers. I knew they were gone, I heard them go.

CONRAD — We had a short discussion about that, as a matter of fact.

GORDON — When I looked in the telescope I couldn't see anything. There was no light or anything coming from there. I thought it must be because I'm not dark-adapted and probably this was correct. The more I stared and the more I looked in there, the less I could see. So, I just went to full range of coverage; full left, full right, up-down, crossways and still couldn't see any stars. I finally went back to zero optics and got the star chart out, and looked at the orbital parameters. The GDC was still operating, so we had an idea of our inertial angles. I looked at the chart and picked up the zero optics point and knew which stars should be coming into the field of view at that particular time.

Fortunately Al was helping me with this. He was looking out his window and could see Orion coming up on his side. So, I just waited until it came into the field of view of the optics. Because of the ORB RATE torquing, the zero optics point was looking actually to the South of Orion. When I looked at my star chart, there wasn't much down there that I could recognize, even under favorable conditions. Fortunately, it happened that Orion did come into view in the very upper-left-hand portion of the optics. When I drove it to the fullest extreme, I saw the belt of Orion dimly in the very edge, and then I could pick up Rigel and Sirius. Once I had picked up Rigel, I could find Sirius. They were the only stars I could see in the entire field of view.

CONRAD — It was a cooperative effort between Al and Dick. Al was looking out his window with one star chart and Dick was hustling in the LEB.

GORDON — The pressure was on and fortunately those two stars were the only ones I ever did recognize. They were Rigel and Sirius. They were just barely in the field of view. I grabbed those two quickly and got a P51 and did a quick P52. I think that one of the stars it came up with was Acamar. I wouldn't have been able to find that without a P52 under any circumstances. But the P52 worked fine and the second P52 over Carnarvon, just before TLI, indicated that we had a good platform. Drift angles were very low. The torquing angles for the second P52 were extremely low and everybody breathed a sigh of relief that we had our platform back again.

5.0 TLI THROUGH S-IVB CLOSEOUT

5.1 TLI BURN

CONRAD — The TLI burn was normal in all respects. My one comment is that we had the little additional program in the CMC to count down to timebase 6 which I thought was extremely good. Dick set it up and had it running. We always knew where we were and the S-IVB performed exactly as advertised. The lunar orbit torquing on our ORB RATE ball worked extremely well. We had very small dispersions about it. I had the decided feeling that, if we had had to fly a manual TLI, it would have gone just like the ones in the simulator and wouldn't have been any big deal.

GORDON — I thought all the cards and procedures that we had onboard were excellent and adequate.

5.2 S-IVB Performance and ECO

CONRAD — The S-IVB performance and engine cutoff were on time and exactly as predicted.

5.3 S-IVB Maneuver to Separation Attitude

CONRAD — The S-IVB maneuvered to the proper SEP attitude.

5.4 S-IVB Maneuver to T&D Attitude

GORDON — We had it all set up. I had the NOUN 17 needles set up for the TD&E attitude and we monitored the S-IVB maneuver with NOUN 17 needles. It was smooth and went right to the preflight attitudes.

5.5 SEPARATION FROM SLA

GORDON — On the separation from the SLA, we were ready in plenty of time. We had everything set up. The separation was made on time with the normal procedures. It was another one of those pyrotechnic events. There was no doubt in your mind that you had come off the SLA and were on your way.

5.6 High Gain Antenna Activation

BEAN — Completely nominal.

GORDON — There were no problems; it was all nominal.

5.7 FORMATION FLIGHT

GORDON — I don't know why we even comment about formation flight. There really wasn't anything to be done at this time. Everything was okay; we had no particular problems.

5.8 TRANSPOSITION

GORDON — I guess my main concern here was fuel usage, procedures, EMS, and the amount of complexity we tend to put into this thing when we're trying to do a very simple operation. We just compound things and procedures by getting fancy and it messed me up once. If I'd been smart and used my head, recognizing all these problems with the EMS in previous flights, I'd have taken these TD&E procedures and scratched all reference to the EMS whatsoever.

CONRAD — Yeah. You have to thrust on time and that's good enough.

GORDON — That's exactly what I would have done. I would have separated from the SLA. I'd have thrust forward on the time, whatever it was, to get 0.8 ft/sec - I'll arbitrarily say 3 to 4 seconds. I'd have waited to 15 and I would have backed off thrusting for a couple more seconds. And that's all you need. This fancy monitoring of the EMS, with these things being bad - there's a hysteresis in them. They go forward before they go down. They're wrong because of the rotational effects and here's what happened to me. The EMS was set up at minus 100 before separation. I hit forward translation and then hit the S-IVB/CSM SEP button and got the pyrotechnic event. I was going to look up to monitor the EMS when I got 100.8 ft/sec and the EMS read minus 98. So I had no idea how much velocity I'd put into the thing and I just continued thrusting forward for a few seconds, probably being conservative because I wanted to make sure I got far enough away from the booster before we did the turnaround. The rest of the procedures are excellent. SCS and the DAP control coming off the booster held that thing right on the money.

CONRAD — You had to do very few up/down, left/right translations. You turned right around and went into AUTO and it was looking right back at it.

GORDON — I used the VERB 63 needles (the NOUN 17 needles came off during the turnaround) and went to SCS control, ACCELERATION COMMAND in pitch, and pitched up at 1½ deg/sec. It was extremely smooth. There were no problems at all. When it pitched 180 degrees, the S-IVB was right smack dab in the middle of the COAS. When I finished the 180-degree-pitch maneuver, I stopped the opening velocity by thrusting towards the S-IVB

AS12-50-7328

and did a VERB 49 maneuver to the docking attitude, at the same time closing all the way. Now I estimate that I was probably a good 100 to 125 feet away from the booster and I was twice as far as I wanted to be. I'm sure the only reason I was there was because of relying on that EMS to give me the velocity when I ought to have kept it as simple as possible. I could have used a clock to get those velocities.

CONRAD — Well, I don't think that distance was too great. That is a big moose back there and you sure don't want to take the chance of running into it turning around. The only other problem is that you want to return and dock. I think that, had we just sat there and left a very low closing velocity and accepted 10 or 15 minutes (which we didn't want to do), we'd have drifted in there okay.

GORDON — Yes. Even at that, I think it was approximately 8 minutes or so before we actually docked and the thing that's disturbing to me is that I got myself in a box and it's due to my own stupidity in not recognizing these bad features. My whole philosophy right now on the EMS, whether it's a good one or a bad one, is that it should be used only two times: One is for entry and the other is for backing up an SPS burn.

CONRAD — Yes, I agree.

GORDON — And when you're doing RCS maneuvers, you can use the G&N or you can do it on time. It's perfectly adequate and it's the simplest approach. The EMS was never designed for anything else other than those two functions. But it was disturbing to me to be able to get in the simulator and consistently dock consistently do the whole thing using less than 20 pounds and then get in flight and end up using 70 pounds. My recommendations now are that the procedures for TD&E are adequate in our checklist, if you scratch all reference to the EMS and substitute times to get your separation velocity.

CONRAD — Yes.

5.9 DOCKING

GORDON — The docking was easy. I have no comments on docking. The spacecraft, as heavy as it is, flies beautifully within DAP control for attitude with the VERB 49 maneuver and it was lined up perfectly with the COAS. All I had to do was translate left, right, up, and down; and control the closing velocity. On contact of the probe and drogue, there were no oscillations, no visible motion at all. I went to FREE as soon as Pete called barber pole and the spacecraft didn't even move. We just sat there for 10 or 15 seconds to assure ourselves that there were no dynamics involved and then we retracted the probe.

CONRAD — Yes, I think the only way you know that you really have two heavy vehicles up there is when you retract the probe. It really is obviously moving a lot of mass together.

GORDON — It takes a long time to get down and then we had the ripple fire on the latches being made and we got the gray indication on both talkbacks.

5.10 CSM Handling Characteristics During T&D

GORDON — All I can say is that the SCS control during the turnaround was excellent. There was no question as to where it was, what it was doing, or where it was going; the control response to the DAP for ATTITUDE HOLD and for the VERB 49 maneuver was excellent.

5.11 Sunlight and CSM Docking Lights

GORDON — My opinion is that we probably never would have any sunlight problems unless we were looking directly at the Sun.

CONRAD — That's my belief too.

GORDON — The docking light was not used, because it wasn't required or needed and that ought to be a closed subject.

CONRAD — The procedures for the LM pressurization went per the checklist. The hatch was removed, the umbilicals were hooked up and CSM power was transferred immediately. We were all aware of our two circuit breakers, which we had to make sure were in and Al had that well in hand.

5.12 EXTRACTION

GORDON — In fact, Al was so concerned about getting that LM out of there that he had both Pete and me check that he, in fact, did have those circuit breakers in. He wasn't going to be the only one responsible. The extraction by spring ejection was smooth as glass. There wasn't any unevenness, oscillation, or any thing. That whole package just whipped right out of there straight as an arrow and we soon lost sight of the S-IVB, as a matter of fact, because it just went right straight aft and was of no concern. We thrusted aft for the 3 seconds that then did the VERB 49 maneuver to the S-IVB viewing attitude.

CONRAD — The maneuver was good; it put the S-IVB right in the center hatch window.

5.13 Photography of Ejection

GORDON — My only comment is that it was probably okay. I had no sense of even having the 16-millimeter running because all I was looking at was the LM. I don't know if Al took any pictures of the ejection or not on his side. I didn't see. You couldn't take any pictures because there was nothing to see.

BEAN — No, I was looking at the TV in the window and you can't do them both.

5.14 Attitude Control and Stability During Separation and Ejection

GORDON — It was good as we mentioned. The VERB 49 maneuver was okay.

5.15 SEPARATION AND EVASIVE MANEUVERS

GORDON — There is no problem here. We were all able to observe the S-IVB. I could see it through window number 2. The maneuver was designed to put the S-IVB in the hatch window or window number 3 for TV purposes and it was a good attitude. There was no particular problem with that at all. We saw a lot of things on the S-IVB venting which Pete called out to the ground at the time. Now here's an area that I'm going to comment about again and I think this is one of the other areas where it got us behind the fuel curve. I don't know whether it was a requirement, but at least tracking the S-IVB during its evasive maneuver was in our procedures. Now, this we did to watch it and the only thing I can say is that it's a bunch of nonsense, for the simple reason that it uses fuel. The less maneuvering you have to do with that CSM/LM combination, the better off you're going to be.

CONRAD — Yes, that's strictly a warm feeling. After that thing has maneuvered to the proper yaw angle and done the APS burn, our observing the LOX dump or worrying about its running into us are wasted effort; Dick's 100 percent right. I guess we shot another 25 or 30 pounds of fuel that weren't in the flight plan trying to track that thing after that LOX dump and we quit before we saw it.

GORDON — Yes, and I just think that it's a very bad thing. We got behind on the fuel curve and I remained behind the rest of the flight because of messing around within the first 10 minutes of flight with the LM docked. All of a sudden, I began to realize that you don't want to do anything with the LM docked except to maneuver to an attitude at 0.2 deg/sec with a wide dead-band, get into PTC as soon as you can, and turn off those thrusters.

CONRAD — If I can remember the numbers right, when we wound up, we didn't use any extra gas after that time. We were some 93 pounds behind in RCS propellant and the rest of your ground calculations of nominal usage were very, very good and I don't think Dick either lost or gained.

GORDON — I don't think I did either.

CONRAD — We must have shot 30 on the extra distance an the docking and the other 60 had to go on trying to hack around to track the S-IVB which was more than adequately far away from us. There was no doubt in my mind that it wasn't going to hit us once it made the APS burn, once it had maneuvered to the proper attitude which we saw it do. That didn't cost us any gas. Once it made the APS burn, it was on its way and wasn't going to bother us again whether it did a LOX blowdown or not.

GORDON — That's right. In defense of some of this fuel usage, I'm not sure that doing the P23's with the LM docked might not have used more fuel than was in the budget. That DELTA in there from separation to the end of that P23 - we shot a lot of fuel and I just can't recommend more strongly that you just flat don't maneuver with that LM on there.

5.16 S-Band Performance

CONRAD — The S-band performance was nominal throughout all that.

5.17 S-IVB Slingshot Maneuver

CONRAD — The S-IVB slingshot we didn't see.

5.19 Workload and Time Line

CONRAD — As far as the workload and time line go, there's more than adequate time throughout that whole operation.

BEAN — Yes. You pointed out in real time that it looked like the S-IVB had some sort of non-nominal venting back there by the engine. It looked as though, right around the upper end of the engine, there was a broken line or something. It was venting and I think from seeing the pictures of the TV that Huntsville's going to want to figure out what happened back there.

CONRAD — One other thing that we didn't mention was that at CSM/S-IVB SEP we had one occurrence that I hadn't heard of before. Normally, it's been the fuel valves. We had a helium 1 B and a fuel secondary A barber pole on us. Of course, we were advised to check these. I watched them. As soon they barber poled, I reset them and that was that.

6.0 TRANSLUNAR COAST

6.1 IMU Realignment and Optics Calibration

GORDON — IMU realignment. There never was any problem in the rest of the flight with any of the IMU realignments. As fax as I'm concerned, you can get any star angle difference you choose to get. If you want to spend the time and make sure that you have the stars exactly in the center of that reticle and the sextant, you can get 00000. If you don't want to be that fastidious with it, you can accept 0.01, or if you just want to put the star in the center and mark on it, you'll get 0.02 and it really makes very little difference as to what you get, other than your own personal satisfaction or gamesmanship.

CONRAD — Yes, let me comment there. There's no doubt that this involves a learning curve because Dick did a couple of 00002, and then after a while he consistently got 00001. We decided that maybe he'd been too rough with the optics and banged them off the stops doing P23's or something; and sure enough a little later, he started getting 00000 and I think that was strictly a learning curve that's not 100-percent obvious to you right away. I think it came with the longer we went through the flight and the more he handled the optics, the better he got at knowing exactly where to put it to get 00000.

GORDON — Well, it's a gamesmanship problem anyway.

CONRAD — If it's 00002 or less, it's in the noise level and we never saw any difference in torquing angles.

GORDON — And you can even do that during PTC if you want to. You can do it at medium speed and keep that star right in the center. There's no problem. So that takes care of the alignments as far as I'm concerned for the rest of the flight. The P23 procedures we used during this flight, I thought, were excellent. There's no problem with them. The VERB 49 maneuver to the optics calibration star was no particular problem. It's a maneuver, however, with that LM out there on the nose that we wasted gas on I'm sure, because I feel now that we could pick any star for the optics CAL. In fact, the P23 star that I used trans-Earth coast was Gienah, which is probably one of the dimmest out there, and it's still adequate for the optics CAL. There's no problem with seeing that star in the landmark line of sight. So, we didn't need to go to the best star in the sky and use 12 or Sirius for that optics CAL. It's completely ridiculous. I recommend that you go find a star that's as close to, or if not at, the VERB 49 attitude as you possibly can to commence the P23's.

I recognize that all the fuel is used in starting and stopping a maneuver, and it doesn't matter how far you go, but it just seems more logical to me that you just take a star as close to the P23 attitude as you need to. I guess the P23's went all right as far as I was concerned. There were no particular difficulties with doing these. I thought the stars and the angles that the ground supplied beforehand were perfectly adequate. I'll say one thing about horizon identifier or locator, whatever you want to call it. That was misleading to me even though I had taken the opportunity to go over to the G&C building and spend a night up there doing P23's off of a slide.

I think this whole problem of marking on the airglow layer for the P23's as opposed to what you might consider the true horizon of the Earth is completely overplayed. It was misleading to me that the first series of marks that I took, I was on a portion of the airglow that was 49 miles away from the true Earth horizon and it was my own fault that I didn't recognize it at the time. I was using a part of the blue airglow layer that appears blue in the landmark line of sight when I should have been using the one much closer to the Earth, which appears as an amber color in the landmark line of sight. This was used during the second series of P23's and gave me a delta H of 19 miles and which was the one I used for all the P23's trans-Earth coast.

I think that whole thing of not marking on the true physical horizon and using an airglow is much overplayed and it was misleading to me. But other than that, the ability to take the marks and to stop the motions of the spacecraft, particularly with the LM docked, and the ability to identify the substellar point and to put the star in the substellar point is an easy task to accomplish. There's no problem with it.

6.2 Systems Anomalies

GORDON — I noted none at this time.

CONRAD — No. Other than the gauges that were out, we noted none. As a matter of fact, I'll make the comment right now. We had no master alarms except for O2 high flow throughout the whole flight. Never saw a thing. And in that respect, CSM 108 was a beautiful spacecraft.

6.3 Modes of Communications

CONRAD — Modes of COMM on the way out were satisfactory. The ground handled the antennas most of the time. We didn't pay much attention to the OMNI switching. One comment I would like to make that happened both in the LM operations and in the CSM operations with respect to COMM, I think we got the ground on the same frequency. It was something that was not apparent doing SIMS. Nor was it apparent until we were in flight. Nor did I hear anybody mention it in any other flight reports. But, there's a decided COMM dropout when they switch stations. We finally got the CAPCOM's up to speed so that 2 or 3 minutes before we got a station handover, especially on the big dishes, they seem to drop out and disappear for 2 or 3 minutes. It never failed to happen at a time when we were communicating with the ground and the next thing you know, we wouldn't be talking to anybody. So they finally got cued up and I recommend, as a standard procedure, that the CAPCOM get a standard callout from the AFD, or whoever is handling that, that there is a station handover coming up at 2 or 3 minutes and just cease COMM until they come up on a new station.

6.4 Passive Thermal Control	CONRAD — Passive thermal control worked extremely well. Tom and those guys worked out good procedures. We had no trouble with it. We lost it one night when we started it poorly. We had a water dump after we started it, or something like that and it was ridiculous. We all should have known better, both us and the ground. And Dick had to restart it one time. That's the only time that happened. Dick has covered the P23's.
6.6 Midcourse Correction	CONRAD — We made our one and only midcourse correction on the way out. It was our first SPS burn. It went exactly like the checklist and, if I remember right, it shut down with 0.1. It was ridiculous, it was so good. No trimming was necessary under the rules and that was either 61 or 62 feet per second and a very smooth burn. We used both ball valves and everything went as advertised.
6.7 Photography, Television	CONRAD — We had our little gauge for the Earth and the Moon when they were full-view in the 500 millimeter and the 250 millimeter and we took pictures on the way out. GORDON — We didn't use a 500 on the way out, I don't think at all, did we, Al? BEAN — Yes we did. We used it a couple of times, some Earth pictures. We tried to handhold it. It will be interesting to see how they came out. CONRAD — Unfortunately we could not get the pictures of Mexico that everybody wanted because of tracking the S-IVB and everything; so, we decided we'd fool around with the TV and we used too much fuel and we just quit. So we didn't get those pictures.
6.8 High Gain Antenna Performance	CONRAD — High gain antenna performance worked fine until later in the mission. We can talk about that. Dick already has and we can get a pretty good handle on what happened.
6.9 Daylight IMU Realign and Star Check	CONRAD — Dick's talked about IMU realignment. It doesn't make any difference whether it's daylight or nighttime, it's a P52; whether you're moving or not moving.
6.10 CM/LM Delta Pressure	CONRAD — We apparently had a very tight LM; so, the ground had us bleed it to get a sufficient oxygen environment in there. I might add that our CSM procedures for getting rid of the nitrogen apparently worked as advertised. We had no need for our second purge. By going the route of dumping the LM and repressurizing from the command module for our ingress, we went into the LM the first day, early.
6.11 LM and Tunnel Pressure	CONRAD — The LM tunnel pressurizing went as advertised, which happened early in the game.
6.12 REMOVAL OF PROBE AND DROGUE	CONRAD — The removal of the probe and drogue was done twice the first day. Once, we went in for the check to see if the lightning had done anything. Then we had a slight argument with the ground because we were sure we left the circuit breakers in the right configuration and they told us we were using too much power. It turned out they were right and we were right. We had a microswitch mis-rig on the upper hatch tunnel which did not turn off the floodlights when the hatch was closed. Therefore we had to pull the flood circuit breaker and we psyched that one out pretty fast.
6.13 ODORS	CONRAD — As far as odors, we did notice a slight odor in the tunnel. It was nothing. It was gone in a short while.

6.14 Passive Thermal Control

GORDON — I think the passive thermal control, the attitude you go to for PTC, whether it is 90 or 270 degrees, should be dependent upon which hemisphere the Earth-Moon system happens to be in. And for the picture purposes going out, we would have been much happier if we had gone to the 270-degree pitch attitude as opposed to the 90, because that would have brought the Earth and the Moon up into the number 1 and 5 windows. As it was, it was down pretty low in it.

CONRAD — If you're going to do a good job on the photography, you should stop. The thought never entered our minds to stop en route for photography because we were behind the power curve on the fuel. If you get off with a batch of fuel and you're not behind the power curve on the fuel, the 4 or 5 pounds it would cost you to start and stop at the proper place to get good Earth-Moon photography en route.

GORDON — I think you can get those pictures even during PTC, provided you select either 270 or 90, depending on which hemisphere the Earth-Moon is in. To us, it didn't make a bit of difference whether we were looking at the south pole or the north pole or which was up or down. That didn't bother anybody.

CONRAD — No, but I got a sneaking suspicion that a handheld 500-millimeter picture is not going to be worth a hoot. We'll have to see when we get back.

69-H-1806

GORDON — I don't think a handheld 500-millimeter is going to be worth a hoot even if you're stopped. The fact that it's handheld is to your detriment.

CONRAD — That's what I mean. If you're going to take that kind of photography, I think you have to stop and point the spacecraft.

GORDON — Well, there is no way you could use a bracket for that 500 millimeter. It points out the rendezvous window with the LM out there. There's no reason to do it. I think it's a good comment to go ahead and select either 90 or 270, dependent upon where the Earth-Moon system is. Well, you can at least look at them every time you go around.

7.0 LOI THROUGH LUNAR MODULE ACTIVATION

7.1 Preparation for LOI Burns

CONRAD — The only thing that I can say is that preparation for the LOI burn went per the checklist.

BEAN — The only thing that I noticed was on the fuel line temperature of the SPS. It was up at 90 degrees at the start of the burn and this is where it had risen after the midcourse correction. The maximum limit is about 75 degrees. We were a little concerned about that. We knew the ground was watching. I also noticed that, as the burn started, the temperature immediately dropped to 75; then, after the burn was completed, it rose again to 90. And this was the situation for all subsequent burns. Another interesting anomaly was that the oxidizer pressure on the SPS was 10 psi higher than the fuel pressure at the start of the burn. And then as the burn began the pressures equalized and stayed there. We never saw this DELTA-P again for a major burn. I guess it must have had something to do with regulator lockup. I might also mention the performance of the PUGS. We monitored the PUGS during the first midcourse correction. After the burn started, I moved the PUGS from NORMAL to FULL INCREASE and left it there for the entire flight. At the beginning of each burn, the

PUGS unbalance meter jumped around a bit as the burn started, then settled down, and at the end of each burn indicated about 100 pounds increase which agreed with the usage; that is, the differences between the oxidizer and the fuel gauge reading itself. So it looked like it was a pretty good move to go ahead and put it in INCREASE and leave it there the whole flight.

GORDON — Of course, that's going to vary depending upon each individual SPS engine. For our particular system, that was a smart way to go; when we got all through with the SPS engine, we were down to 50 pounds increase and that was it.

BEAN — You're right. They were able to predict it fairly well for our engine and if they can do it for others, the crews will perhaps be able to do the same sort of thing with another engine. It worked well for us though. It saved having to fool around with that during the burn.

GORDON — If we had fooled around with it, we'd have never caught up with it, because one time it did go to DECREASE, I think, and if you had gone there, you'd have never caught up.

GORDON — The burns were nominal; there is nothing to say about them.

7.2 SPS BURN FOR LOI I AND LOI 2

CONRAD — As a matter of fact, the engine shut off 6 seconds early on LOI I and LOI 2 was too brief a burn to shut off on time. I think the ground had a better handle on the engine for time anyhow. Everything went by the checklist. We went right by the LOI I abort rules. We called out "tight," "loose," and "tight." I followed the stuff and monitored the burn to determine whether we were in tight or loose rules. And I'd like to mention that we got fairly complicated on mode I, I-A, II-A, II, and all that business and really there were too many of those to call out in the cockpit.

CONRAD — So we went on the tight/loose scheme. Whenever we were in the tight rules, we called "tight"; when we were in loose rules, we called "loose"; and then we went back to "tight" and in that way we let the modes fall out where they would....

GORDON — Where they would if we shut down.

CONRAD — ....if we shut down. And all that I-A and II-A business; whether it was APS/DPS or DPS only or anything like that; whether it was a 2-hour or 30-minute one; that got kind of complicated. They were simple when we first started out about 4 months ago, but it got exercised into a very complicated thing. I think it's a good exercise though, because we had a lot of confidence before we went that we could burn both the DPS and the APS to get ourselves out of trouble. And so in that sense the whole thing was a good exercise.

7.4 Maneuver for AOT Star Observations

CONRAD — We were in a position, when we entered the LM before LOI and when we entered it after LOI, such that Al was satisfied that he had seen enough through the AOT.

BEAN — Well, we saw Dick.

GORDON — Out one of the hatch windows.

CONRAD — Yes, he can look right in the command module if you put it in the back detent.

GORDON — Yes. I can look back and see their eyeball.

CONRAD — LM communications were as advertised when we powered up.

CONRAD — The television worked in lunar orbit as well as anyplace else.

CONRAD — The tunnel mechanics and pre transfer operations were straightforward. We had a set of checklist items that we packed and took over at the various times.

GORDON — IVT was no problem. We were all over there at one time or another.

CONRAD — Yes. Everyone whistled in and out. I don't know about anybody else, but I never got disoriented going back and forth. It seemed relatively easy to me to straighten out where I was going. I could go in and out head first or feet first, in either direction, and it didn't make any difference.

CONRAD — That was in good shape. We mentioned the upper hatch micro switch. That was the only anomaly that we noted.

CONRAD — I think that the LM stowage is excellent. It's well worked out and well thought out; everything worked and we had no problems.

CONRAD — Power transfer was nothing.

GORDON — I don't know why we even talk about that. RESET, OFF; not too difficult.

GORDON — I really don't think there's much to comment about. The procedure is established. The attitudes, the 0.3 deg/sec pitch rates, were easy to perform. You need a good map, a good photo of what you're trying to look at. And when you look through the telescope the landmarks ought to be easy to recognize. The AUTO OPTICS feature of P22 puts you so close to the actual landmark that there is just no doubt about it, but it's reassuring to have a good photograph of the landmark and to know exactly what you're going after. The LM does occlude the target for a considerable time and I think the ground has already worked out those attitudes to make this an easy task. All I can say is that it was easy. There is no question about anybody's ability to do that. I think it's one of the easier things and I was surprised that it was as easy as it was from the experiences that we have had in the CMS. I have a recommendation here in that particular regard. They ought to provide slides of the landmarks that the crews are going to use on their flights. We have pictures of them and, because we have pictures, we could make slides for use in training - slides of the exact landmarks you're going to be looking at in lunar orbit. That really didn't make much difference, but it gives you a little warmer feeling as to what you're going to be looking at.

GORDON — The same remarks apply as going the other way. It's a two-way street to that tunnel.

GORDON — This is the CSM power transfer to LM at 104 hours; it was done on time and on call from the LMP. It was done without any particular problem, as expected. The time interval between that power transfer until closeout, of course, helped Pete finish putting on his PGA, making sure the zippers were all locked out and secured.

8.1.2 Tunnel Closeout

GORDON — Tunnel closeout was actually accomplished in the suit with helmet and gloves on. There was no particular problem. We had plenty of practice with that tunnel, during transfer and coast. We were into the LM on a couple of occasions. We had already had a chance to look at the tunnel hardware, and it all performed and operated properly. Pete put the drogue back in the tunnel; I installed the probe, hooked up the umbilicals, preloaded it, and then cocked all the docking latches. All of them cocked with two strokes with the exception of the three that were called out during translunar coast as being only partially expanded. They weren't all the way in and it was expected it would only take one stroke to cock those two latches. That was the case in fact. There was no surprise and everything went according to plan. The hatch was put in without any difficulty whatsoever. When locked and sealed, the tunnel bled down to approximately 2 psid where it was stopped. The integrity check of the command module was conducted at that time and there were no leaks. The tunnel was bled down the rest of the way. It was a smooth operation. As soon as that hatch was in place and secured and the integrity check completed, I removed my helmet and gloves and stowed them.

8.1.4 Undocking

GORDON — The undocking, from the maneuver to the undocking attitude, was done using the P30 and P41 from the PAD values sent up from the ground previously. That went without a hitch to DAP maneuvering, and no particular problem was experienced. We could have been in the simulator for all I was concerned. The soft undocking was done on time. The EXTEND switch was hit just momentarily. I guess I was a little surprised at the speed at which that probe actually extended and at the reaction when it hit the end. I suspect, in watching it, there was a little rebound when the LM hit the end of that probe. There was nothing large or exciting, but it was a rebound that I really hadn't quite expected. Neither vehicle showed any tendency to diverge or oscillate or anything; everything was quite stable. I sat there for probably 30 seconds looking at the vehicles to make sure that they didn't oscillate or have any perturbations. Then while I was holding the probe switch to the EXTEND position, prior to aft translation - for about 2 seconds - the vehicles

69-H-1999

parted cleanly without disturbing either vehicle's attitude. The command module was in SCS ATTITUDE HOLD at the time with RATE, LOW, and DEADBAND, MIN, and the DAP had been previously configured to CSM ONLY.

As soon as we separated, I switched the probe back to RETRACT; and, without really trying to station-keep at that very minimum distance, I let that separation velocity imparted to the two vehicles just translate the CSM away from. the LM. It just slowly drifted back until I suspect we had probably 100 to 150 feet, somewhere in that vicinity, at separation time. It was smooth. Nothing unusual occurred other than that rebound that I did notice when the probe hit the LM.

8.1.5 Attitude Control Modes

GORDON — It was a little bit different seeing the LM no longer on the nose after 3 days or some 104 hours. But as far as the control modes are concerned, the DAP provided an excellent control mode with no particular problems at all.

8.1.6 Undocking and Separation Photography

GORDON — The undocking and separation photography, sequential and Hasselblad, was conducted without any particular problems. Hopefully, we got some good photography from this whole sequence. In fact, the television was on during this time, and I had the monitor down between my legs. It was on the extendable storage compartment, right above the number 1 couch, right above the seat belt right between my feet. I could actually see the pictures that the TV was taking and occasionally, I reached over and adjusted the zoom control as well as the focal length. The TV was mounted in the right-hand rendezvous window.

8.1.7 Formation Flight

GORDON — There really isn't much to be said about that. The control modes were with the DAP set for attitude control; I used the translational hand controller sparingly, if any at all, to keep the LM in sight within the field of view of me, the 16-millimeter DAC, and the television camera.

8.1.8 LM Inspection Photography

GORDON — We really hadn't done any inspection of the LM. We were well assured that all four gears were down and locked. I could see three gears before separation, so I knew they were out and extended, so we decided that a LM inspection was not required or necessary at this time. Photography was taken in conjunction with the undocking and separation photography. It was kind of lumped all into one big package. The separation maneuver was simple in itself. I might stop here just before I mention that, Pete had mentioned in his section on the LM debriefing that there was some out-of-plane. I noticed this also, that the LM drifted slightly to my left, but after I drifted off to the right, a small distance, it seemed to stop and stabilize and there wasn't any particular concern. I didn't waste the fuel trying to get back in plane for the separation maneuver. I did another P30/P41 separation attitude; and, as would be expected with a 90-degree orbital travel, there was no, or hardly any, attitude change at all. It was more or less a re-trend maneuver than anything else, or a trend maneuver, until the separation attitude. That's the point in time a 2.5-ft/sec Z-translation was performed and we were separated.

8.1.10 Rendezvous Radar and Optics Check

GORDON — All I can say is that they worked okay. My VHF corresponded with Pete's rendezvous radar and, at that particular time, there was no concern about the optics checks. In fact, I waited for quite a while before I even went to the P20 attitude to take a look at the LM through the optics.

8.2 LUNAR MODULE

BEAN — We ingressed the LM exactly on time. The only difference between the checklist and what we actually did is, both Pete and I suited up before we got in the LM. I think this is a good plan. This way, we were both able to listen all the time to what was going on during the LM checkout, and also, didn't have any unusual circumstances come up when I was supposed to go back into the LM and put on my suit.

CONRAD — We didn't want to make a change to the flight plan because we weren't sure whether it would work or not. Listening to previous occurrences, some people had problems suiting and some did not. Now all three of us are small, and we all three managed to do the suiting exercise in the amount of time laid out for one person. All three of us had no difficulties getting in the suits by ourselves. The only help we gave to each other was to actuate each other's zippers.

BEAN — Just as we were getting suited up, the doctors came up with a comment that my biomedical harness wasn't working. So, we had to stop and spend about 15 or 20 minutes replacing sensors. My only comment here is that they ought to have a handle on these sensors. By the time you finally get to the Moon, if it looks like one of them's going to be going bad about that time. Maybe we ought to, a day previous to that, go through the sensors and find the one that's bad and fix it. Actually, we had to take off three of them before we found the bad one. That was at a bad time to go through that exercise. We could have done that the night before. No use waiting until the last minute.

8.2.1 Power Transfer Activation and Checkout

BEAN — Everything went exactly as planned. The transfer took place at 104:00:00. All the caution lights, warning lights, and operational checklists were completely normal. The voltage on the batteries was a little bit lower than would normally be the case; we had one low voltage tap, because we'd gone in and activated the LM two previous times, but this didn't give us any problem at all. We just went over the high taps immediately and the rest of the checkout progressed normally.

8.2.2 ECS and Suit Loop (Cabin Atmosphere)

BEAN — ECS and suit loop operations were normal all the time. The only difference we encountered was the suit integrity check that was performed after separation and prior to descent. This normally is done at 3.7 pounds above cabin pressure. We didn't do that. We did that only at about 2.5 pounds above cabin pressure, because my ears were stopped up at that time, and I was unable to clear them.

8.2.3 VHF & S-Band Communications; Steerable & OMNI Antennas

BEAN — They were all excellent; operated just exactly as we had hoped.

8.2.4 PGNS Activation and Self-Test

CONRAD — We got two alarms, neither of which was expected. Although I closed the ground, I wasn't too concerned about them. If I remember correctly, we came out of STANDBY and brought the computer on the line. We got an alarm, an 1100 series alarm, which was "uplink too fast." Then in doing VERB 35 (Warning Light Check and DSKY Light Check), we got another alarm which said "PIPA fail." But PIPA is not in use, which was a 212 alarm if I remember correctly. Those were the only two anomalies. Everything else went exactly as per activation checklist.

8.2.5 AGS Activation, Self-Test, Calibration, and Alignment

BEAN — This went per the checklist. We also looked at some erasable memory locations that were voiced up the previous day from the ground to see if they had been affected at all by the lightning strikes we had at launch; they were not. All the numbers were exactly per the data that had been voiced up. The only thing I noticed that was slightly anomalous was when we performed PGNS to AGS alignment - when I would check the FDAI, by moving the attitude switch from PGNS to AGS, the ball would jump about a quarter of a degree and sort of roll and pitch. The ball would jump. At other times, when I would make that alignment, it would remain perfectly still, indicating, I guess, a perfect alignment. On all these cases, we did a VERB 40 NOUN 20 beforehand, and I never did figure out exactly why it didn't make a perfect alignment each time.

8.2.6 Ordeal

CONRAD — The ORDEAL worked as advertised.

8.2.7 Deployment of Landing Gear

CONRAD — Deployment of the landing gear left no doubt in our minds that the pyros fired. Dick was able to see three of the four gears from the command module and we got a gray talkback, which indicated the gears were down and locked. Everybody was satisfied that the gears were down and locked. Therefore, we proceeded as planned on the undocking and we did no turnaround for Dick to inspect downlock.

8.2.8 DAP Loads

CONRAD — The DAP loads went as they were listed in the checklist and updated from the ground.

8.2.9 Rendezvous Radar Landing Radar Self-Test

CONRAD — We had two minor anomalies on the rendezvous radar self-check. Our first anomaly on the rendezvous radar check was that we did not get 500 feet as advertised on our tapemeter. We got 493 feet. The systems specification number was 500 feet, and 493 feet were specified for our particular radar. The check list did not reflect our specific numbers. This is something that we argued about before flight, that we did not want a specification number in there. We wanted our right numbers in there. Somebody chose to disagree with us and continued with the specification numbers. The other anomaly, although it didn't affect us because we were never at great ranges, was that our power was slightly low. It should have been on the order of 3.7 volts. It was 2.65 volts, and it remained that way throughout the flight. We achieved lock-on, both on the lunar surface and in flight, at some 240 miles or so with no problem at all. The rest of the checkout, the four gimbal angle readouts, the platform alignment, and everything went per checklist.

The one thing that we did was that anytime we could get ahead on the checklist by taking whole blocks at a time, we would go ahead and do these early. Now our very last SIM before we left Houston, we ran a SIM which came out almost exactly the same as in flight, in that we pressurized the RCS early; we got our torquing angles back up from the ground before we lost MSFN on the first pass, instead of having to wait until AOS on the second pass.

This allowed us to get many things done in advance. Dick got the tunnel closed out early, we had the LM closed out early, and we were in the independent mode still docked well in advance. We had our RCS pressurized in advance. As a matter of fact, I think we got that done far enough in advance that the ground was able to observe RCS pressure, which was not a requirement. Therefore, when we came up on the second pass, we managed to get our really tight one out of the way early, before Dick did his tracking and the RCS checkout, hot and cold fire. This would not have worked as smoothly had we not practiced this with the ground, and the ground was anticipating our hustling.

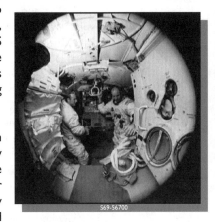
S69-56700

You can't do enough of these SIMS with Houston so that Houston gets used to the individual crew and how they operate. We let them know that we were going to hustle, they were anticipating our hustling, and they were ready for us. This really worked well as far as the gimbal angles and getting things done in advance were concerned. Had we run into a problem somewhere along the line, we would have had more than adequate time to cope with it. Fortunately, we didn't run into any problems, so we had a lot of dead time, which is the way it ought to be.

8.2.10 Undocking

CONRAD — Now, we were all checked out and ready to go and sitting around waiting for undocking. The soft undocking worked very well. When Dick undocked, we hit the end of the probe, and there were some very slight longitudinal oscillations; but the probe damped it very well. When he undid the capture latches, the two space craft were completely null to each other and did not separate. Dick physically had to back off. I had no indications from either the AGS or the PGNS that we got any velocities, if I remember correctly. What we really got was some slight PIPA bias, in the order of 0.1 ft/sec, which was well within anything that we were going to null. So, we never nulled anything in separation.

8.2.11 Formation Flying

CONRAD — Formation flying worked very well. I did the yaw and pitch maneuver, and after that Dick did nothing but slowly drift away from us. We were so well stabilized at undocking that the little SEP maneuver that Dick did to back away from us at undocking was about the only time he touched anything. We remained in good sight of each other all the time. Neither spacecraft imparted velocity to the other or to itself.

<table>
<tr><td>

8.2.12
SEPARATION

</td><td>

CONRAD — Dick performed the SEP maneuver, and for some reason it looked out of plane to me. I don't know why. They may have gotten a slight amount of out of plane, but very small. It did look like he drifted a little bit out of plane to our left as we faced him, which would have been to his right. At that point, we busied ourselves with the time-line portion of our checklist, which was to get into the landing radar check and, as soon as that was done, to perform our first alignment in the darkness.

</td></tr>
</table>

8.2.13 Lunar
Landmark
Recognition

CONRAD — Lunar landmark recognition was easy. MSFN relay was not used by the LM.

BEAN — When Dick was doing his maneuvering to the landmark track attitude, he had the S-band antenna in AUTO TRACK. Sure enough, just like they said prior to flight, when he fired his thrusters, I could hear the antenna move. It would come back to its AUTO track position. I had forgotten to put it in either the OMNI or the SLEW position, which I think is a lot better than AUTO track; because, when he drives it off with his thrusters, it's likely to lock up on the side lobe or something like that.

BEAN — So, I would recommend next time that we either go to OMNI during that period or to SLEW. The antenna makes a lot of noise, and you can hear it move around. The medical kit in the LM doesn't contain any Afrin and it ought to. When you really want to have the ability to clear your ears, if you're a bit stuffy, it's before you go EVA. With no Afrin, you're going to be stuck. We had taken some over from the command module. I noticed that you need to use the restraints in the LM if you don't have on your pressure suit. If not in the suit, you didn't really need to attach any restraints; you could float about and do whatever you had to do. But in the suit, you need to attach a restraint. The TSB over on Pete's side - the lower loop on the back side is supposed to hold it down. It does not perform that function because the dimensional relationship between the place that it hooks in and the place that it hooks to the LM allows the TSB to float up, and it tends to bother you. That ought to be rearranged so it holds it down. The K-factor update from the ground came up in about 1 minute after we updated the AGS. That's the way it ought to be. We never touched it after that. The rate needles had the same problem that they've had on all previous flights; i.e., when we actually had zero pitch, Pete's was reading minus 0.5 deg/sec, whereas mine was reading plus 0.5. Did that ever bother you?

CONRAD — No, because the only time that I flew manually, which was after ascent, I had radar lock and I never used the rate needles. There was no reason to. We did all our AOT work by AUTO maneuvering the star; and then I flew in PULSE again and used the attitude error needles to indicate what I was doing. As soon as we undocked, we got on the SEP checklist; and that went exactly the way the checklist was written. There were no anomalies; everything worked as advertised. The ball angles were good. The landing radar self-check went exactly as advertised. The numbers, which I believe were our radar numbers, came in perfectly. We knew we had a very good landing radar. We went all the way through to SEP, and Al got some pictures of CSM SEP. Unfortunately, that's the magazine we left on the lunar surface. SEP was at 108:24:22, and the time line had it published as 108:24:22. So we were in good shape, right on to nominal.

BEAN — The changes we made in the PAD, so that they were shorter than those on Apollo 11, worked out real well. I don't think we had a bit of trouble getting the information or copying it. It didn't interfere at all with all the other things that we had to do. So I think the PADs are in pretty good shape.

CONRAD — We were passed the proper torquing angles that were to meet our limits, which were quite wide anyhow. Our torquing angles were very low on our alignment.

AI and I had practiced doing a two man alignment. AI looked at the stars. I flew the vehicle and ran the computer, and we got the alignment done in pretty snappy order. I think AI came up with 00001 on his first alignment. Is that right?

BEAN — I can't remember.

CONRAD — They were all good.

BEAN — Yes. I think that's the way it ought to be done. That's a good way to do the alignment. It lets the fellow who's looking out at the stars keep his night vision up. He doesn't have to take his eyes out of the AOT. Pete was down there using the attitude error needles, which worked real well, and punching the computer. We were able to do it very rapidly without any errors. There weren't many changes passed from computer to AOT to computer. Everybody had something to do, and it went very rapidly - the same way on the lunar surface and the same way after ascent.

CONRAD — Yes. We worked three marks for a start, and that seemed to give us very good alignment. Our rendezvous radar check, which came before the alignment, went as advertised. The tapemeter checked out. The needles checked out, and the PGNS readouts in VERB 63 checked out. We were very happy. The stars that we had picked for the alignment, Capella and Rigel, were good stars. We had no difficulty finding them. The AUTO maneuvers were smooth. As soon as we had done that, we went to DOI attitude. We went through our checklist and set up for DOI, and did it exactly from the checklist on the time line.

9.0 DOI THROUGH TOUCHDOWN

9.1 Command Module

GORDON — There was no particular problem with this one. I had the P76 PAD ready to go in the CMC for updating the LM state vector. As a matter of fact, I watched the DOI burn through the optics and it was kind of funny to see that engine burning right at you during the burn. Through the sextant it was almost like looking right straight up the tailpipe of an airplane. That object was out there and I was looking right down the descent engine nozzle, watching the burn. Through sextant, of course, it looked like it was right next to me. We were only 3.4 - 3.5 miles apart during that time.

9.1.3 Optics Track - Ease of Tracking LM

GORDON — There was no particular problem here and as far as the optics, ability to look at the LM; I could read it through a telescope or the sextant. It was easily visible and tracking with both optics was exceptionally smooth. I guess you might say it was a pleasant surprise, from the experience that most of us have had in the simulators - that the optics drive in the CSM was exceptionally smooth. It was really a simple matter of being able to follow the LM throughout this whole phase.

9.1.4 SXT/VHF Track

GORDON — All this after DOI was no particular problem. I put in a P76. Preferred tracking axis was verified right on the LM and there was no particular problem. The VHF ranging was good. It stayed locked on. There was no particular problem and the sextant, of course, was tracking the LM. Now, something happened here - I had planned to follow the LM down from DOI to PDI using the sextant and VHF tracking to update their state vector. For some reason, I was doing something with the computer and at the time there was a NOUN 39 right after the P76 was inserted on VHF ranging. It was a fairly significant one, rather large, and without even thinking or recognizing what it was at the time, I was doing something other than worrying about tracking of the computer. I looked up there and saw those numbers and I just proceeded to get back in a normal P20 sequence and I'll be darned if I didn't proceed with a very bad NOUN 49, which I should have rejected. It was the first one that came in. And, of course, the computer accepted it and it kind of blew the state vector. I looked out the sextant and saw the

LM again, but then I really wasn't very concerned about keeping the LM state vector up to date. In fact, by accepting that bad VHF, the LM state vector was essentially lost; so, rather than be concerned about tracking the LM from DOI to PDI, then as far as I could for training, I just forgot about it.

CONRAD — When did you put that in?

GORDON — Right after DOI.

CONRAD — The first mark?

GORDON — Yes.

CONRAD — How bad was it?

GORDON — It was several hundred miles or something. It was just out to lunch; it was just nothing there. I guess it was taken at the time the VHF was coming in. I should have known beforehand not to accept those particular marks. Without thinking, I just reached up and hit the PROCEED button, without recognizing there was a NOUN 49 sitting there.

9.2 Lunar Module

9.2.2 DPS/DOI Burn and Performance

CONRAD — The DOI burn was excellent. The burn was on time. The residuals were zero, plus 0.2, and minus 0.6. The AGS residuals were plus 0.3, plus 0.1, and minus 0.6. Because we had very good agreement, we didn't exercise the rendezvous radar option and make a lock-on with the command module. We pressed on. Our computer showed that it put us in 60.5 by 8.9, which was very close to the pads I used for DOI.

BEAN — The AGS showed 60.5 by 8.5.

CONRAD — We immediately went to the PDI attitude, 01090, which left us facedown looking at the Moon. From that inertial attitude we watched ourselves pass from face-down through local horizontal to pitch-up at PDI. It gave us an excellent look at the Moon going around; and we had a relatively easy checklist at that point. We reset the DAP and our 20-degree stick and accomplished the checklist, put on the helmets and gloves, buttoned everything down, put the COAS in the overhead window, and went right down the line. We had lots of time. We brought on the ascent batteries. Did we bring them on early that time? The ground wanted us to bring them on. One time they wanted us to bring them on early. I guess battery 5 was a little low. It took a little longer to heat up later. Now, we got to our perilune and altitude checks and that's a place that we didn't realize any error could crop up. It turns out that the predicted perilune and altitude checks at PDI are based on the command module being in some fixed orbit. It turns out that Dick was not in that orbit. We realized something was wrong with the checks because, not only did it show us high, but it didn't show a consistency. At each mark, our altitude at PDI grew from 56,000 feet finally up to 64,000 feet; and the ground said, "Forget it," because they had a good track on us and said that we were still showing a 50,000-foot perilune - right on the money.

BEAN — I'd like to recommend that this check be eliminated. It's just busy work. As you can see, it didn't provide any intelligence that let us make a decision. No matter what the numbers would have come out on that, I think we would have gone ahead and made the landing just as we did because it showed us diverging from our nominal PDI altitude and headed up. So, unless we come up with something better, which I don't think we need with the way the PGNS is working and with good ground tracking, I think we ought to eliminate it. This will give you more time to prepare for PDI; to look out the window and make sure all is in order.

9.2.7 VHF Ranging

CONRAD — We were on the VHF ranging mode for the command module. We could tell when he locked up. We could hear the tones. It never bothered us, and I'm sure that Dick got the information that he wanted.

9.2.8 MSFN Acquisition Via PCM High and Update PADs

CONRAD — The communications were excellent. The PAD updates were given on time, and as Al already commented, were easy to copy. We never had any trouble with telemetry that I know of. We had good telemetry all the way, I think.

BEAN — I've got one comment. I noticed, when we were coming into PDI, that there was a lot of background noise. I guess it must have been on the S-band. It had sort of a roaring, whistling sound, and it persisted for about 10 or 15 minutes. I don't exactly remember when it stopped, but it finally ceased so that from PDI through landing, we didn't have that sound.

CONRAD — Yes, I remember that now. It didn't really interfere with the COMM because the voices were good and loud. This whistling, windy sound was on there.

9.2.9 PDI BURN; PGNS PERFORMANCE AND PROCEDURES

CONRAD — We got a GO for PDI, went into P63 and had our final trims brought on the landing radar. We had ignition on time and the engine was very smooth. It throttled up right on time. We started down the trajectory and, because I could not see the ground, I religiously kept my head in the cockpit and used the ENTER button at 30 seconds, 1 minute, 1 minute and 30 seconds, and right on down the line, carefully checking the PGNS first as to predicted values, while Al checked the AGS versus the PGNS. The AGS and PGNS stayed in very, very good agreement. We got the ED batteries checkout. They came up with a NOUN 69 at plus 4600 feet, if I remember correctly.

BEAN — 4200 feet.

CONRAD — Plus 4200 feet at about a minute and a half. We entered it. The ground read it so that it was okay. We pushed it into the computer. This seemed to be the right number later on because it looked like we were targeted right dead smack in the center of the crater. The landing radar came in exactly as predicted at 41,000 feet. Here, I noticed something that is different from the simulator. The altitude beam locked on and apparently the velocity beam locked on also. Now, in the simulator, if I remember correctly, the velocity light will not go out until V_1 is 2000 feet per second or less. Apparently, the lock-on light is independent of the 2000 feet; that is, when it will take velocity information. The altitude light went out and shortly thereafter at a higher velocity than 2000 feet, the velocity light went out indicating a lock-on. Now, I'm sure that the computer did not use this information. So we got to a V_1 of 2000 feet, which was different from what we had seen in the simulator. That's the only difference that I noted between flight and simulator during P63. The other comment I have in P63, and it didn't bother me too much although I commented to Al about it several times, was the fact that we had considerably more RCS activity than we had noted in the simulator; but Neil also mentioned that, all the way down, he had more RCS activity than he had noted. So, I think this is due to radar updating. Our DELTA-H, the first time I noticed it, was a minus 1100 feet and it had some noise on the radar. It jumped to 1900 feet and steadied out between 1100 and 1900 feet. We watched it for a few seconds, and the ground gave us a good update which we did. I'm sure that we were taking radar data at some 39,000 feet.

9.2.11 BRACKING

CONRAD — The throttle-down did not happen quite as predicted. It seems to me we were off by a second. I don't remember whether it throttled down a second early or a second late. I believe it throttled down a second early. I think they gave us a throttle-down at 6 plus 22 and we throttled down at 6 plus 23 or 24 or something like that. But it was close enough that we felt that that part of it was going all right. Every once in a while, I took a peek to see if I could see the horizon; and, along around 25,000 feet, I could see the horizon out of the bottom of the window, but seeing it didn't help so I went back on the gauges. I'll have to go back to the time period of our alignment.

9.2.12 LPD ATTITUDES AND ACCURACY (CALIBRATION)

CONRAD — We used the COAS option and pumped in a 40-degree angle for the LPD. It was easy to read the LPD in darkness and see where the star was. We had a reading of 0.8 degree right and 0.2 degree down, which was where the star was located. This showed a lot smaller yaw bias than we thought we had. I thought that the numbers were insignificant, and in the noise level for me to worry about them. I was going to interpret the LPD just the way I saw it. We went into P64 at 7000 feet, and I think we hit high gate right on the money.

9.2.13 Final Approach

CONRAD — As soon as we started the pitch maneuver, I proceeded on the computer to enable LPD and immediately went outside the window. For the first couple of seconds, I had no recognition of where we were although the visibility was excellent. It was almost like a black-and-white painting. The shadows were extremely black, illustrating the craters and all of the sudden, when I oriented myself down about the 40-degree line in the LPD, our five-crater chain and the Snowman stood out like a sore thumb. I started asking Al right away for LPD angles; as best as I could tell, we had absolutely zero out-of-plane error. We were targeted right dead smack in the middle of the Surveyor crater and I just left it alone.

I didn't LPD for quite a while until we got down around, as I remember, 2300 feet or so. I LPD'd one right to move it off the crater and headed for the landing area short of the crater. At that point, I listened to some more LPD angles from Al and I had the feeling that I was a little high; so I LPD'd two clicks short, and I let it go for a while. Then I decided that I was going to land a little short and Al called out something like "30 seconds worth of LPD remaining." I gave her one click forward, let her go for a while, and decided we were high and fast. I didn't like the size of the area short, where we had normally been trying to land, and I looked for a more suitable place. At the same time I took over manually at about 700 feet and immediately killed the rate of descent. It looked like we were going at the ground like a bullet. I had plenty of gas and I wanted enough time to look around. At that point, Al got a little nervous because I had killed the rate of descent to 3 feet a second at 500 feet. I left a very high pitch angle on it, on the order of 30 degrees, because we were moving quite fast and I wanted to get stopped. I had the horizontal velocity under control about the time I passed the near edge of the Surveyor crater.

I saw a suitable landing area between the Surveyor crater and Head crater, which now meant I had to maneuver to my left and sort of fly around the side of the crater, which I started to do. I guess I wheeled it pretty hard, because Al commented a couple of times that I was really cranking her around and I told him it was no problem. I had everything under control and I did increase the rate of descent after he called my attention to the fact that we had leveled off quite high at 500 feet, I got down as soon as I got over the area that I wanted to land on. To me, it looked like a perfectly smooth, good area, between Head crater and Surveyor crater and I started a vertical descent from a relatively high altitude, 300 feet at least. It may turn out that I actually backed up a little bit; but I don't think so.

As soon as I got the vehicle stopped in horizontal velocity at 300 feet, we picked up a tremendous amount of dust; much more so than I expected. I could see the boulders through the dust, but the dust went as far as I could see in any direction and completely obliterated craters and anything else. All I knew was there was ground underneath that dust. I had no problem with the dust, determining horizontal or lateral velocities, but I couldn't tell what was underneath me. I knew I was in a generally good area and I was just going to have to bite the bullet and land, because I couldn't tell whether there was

a crater down there or not. We came down with a relatively low descent rate. I think I speeded up to about 6 ft/sec and got her down around 100 feet, where Al called it, and I slowed to about 3 ft/sec and started milking her down.

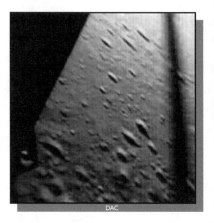
DAC

At that point, the dust was bad enough and I could obtain absolutely no attitude reference by looking at the horizon and the LM. I had to use the 8-ball. I had attitude excursions in pitch of plus 10 and minus 10, which happened while I was looking out the window making sure that the lateral and horizontal velocities were still nulled. I would allow the attitude of the vehicle to change by plus or minus 10 degrees in pitch and

not be aware of it, and I had to go back in the cockpit and keep re-leveling the attitude of the vehicle on the 8-ball. I was on the gauges in the cockpit doing that at the time the LUNAR CONTACT light came on. I had that much confidence in the gauges. I was sure we were in a relatively smooth area. I had my head in the cockpit when the LUNAR CONTACT light came on and I instinctively hit the STOP button and that's how we got a shutoff in the air. We were, I'd estimate, 2 or 3 feet in the air still when I shut down the engine and it dropped right on in.

We landed on a slight slope; therefore, the right plus-Y gearpad touched first and tipped the vehicle to my left. The vehicle plopped down on all four gears at that point with no skid marks that we could determine other than the first pad touchdown. When we set it in for a landing and looked around, it turned out there were more craters around

there than we realized, either because we didn't look before the dust started or because the dust obscured them. One thing that I wanted, and we still haven't figured out why I didn't get it, was the crosspointers. I reached up and hit the switch from HIGH MULTIPLE to LOW MULTIPLE and still had no crosspointers. I asked Al if possibly one of the switches was in AGS, but he said it was in the PGNS position. We should have had cross pointers, but, for some reason, we did not. It's an anomaly that I couldn't check out afterwards. The crosspointers worked with the rendezvous radar, so it's not that my gauge was out. I don't know the reason for not having the crosspointers, but I

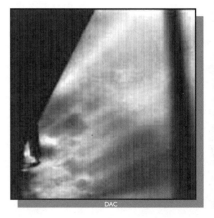
DAC

wanted them very much. I think had I had them I probably would never have looked at the ground in the last 50 or 100 feet. I would have gone completely on the gauges.

Now, the landmark visibility I have to class as excellent. It was very easy from our model and the photographs that we had. There was no doubt in my mind when we finally zeroed in at about 6000 feet exactly where I was, that I was in the right place. Had I been short or long, I think it would have been very obvious to me. The lighting was excellent. The surface had a white-gray appearance; the shadows completely black. I was not aware of any washout and I was not aware of seeing zero phase. Everything looked a lot smaller and closer together in the air than it turned out to be on the ground. When we were on the ground, things that were far away looked a lot closer than they really were.

The color does change with look angles. My yaw only changed maybe 10 or 15 degrees in either direction, both times to the left and right. This didn't make too much difference to me in the change of color. Whenever we looked directly to our right or left, cross-Sun, things had a browner appearance than normal, which was expected from looking at Neil's photographs. None of these things appeared unusual. I think I might have used the LPD more, closer in, but I was sure that we were so much on target that I really didn't need it. That may have caused my slight over flying and taking over to stop what I thought was a relatively high horizontal velocity. As it turned out, I don't think we landed more than 400 to 500 feet past the Surveyor. We were right on the very edge of the Surveyor crater, the far edge of it. I could see it from looking around the side of my window as we sat on the lunar surface.

9.2.14 MANUAL CONTROL

CONRAD — I think the manual control of the LM is excellent. The LLTV is an excellent training vehicle for the final phases. I think it's almost essential. I feel it really gave me the confidence that I needed. I think the simulator did an excellent job in manual control and LPD training all the way down to the last couple of hundred feet. I think both devices worked very well together.

DAC

9.2.15 HOVERING

CONRAD — Hovering was easy.

9.2.16 BLOWING DUST

CONRAD — I've already commented on the blowing dust. I felt it was very bad. It looked a lot worse to me than it did in the movies I saw of Neil's landing. I'm going to have to wait and see our movies to determine if it doesn't show up as badly in the movies as it does to the eye. Maybe we landed in an area that had more surface dust and we actually got more dust at landing. It seemed to me that we got the dust much higher than Neil indicated. It could be because we were in a hover, higher up coming down; I don't know. But we had dust from - I think I called it around 300 feet.

9.2.17 TOUCHDOWN

CONRAD — I mentioned the engine off while still airborne. That pretty well covers it through touchdown.

BEAN — We gave them two or three AGS updates for altitude every time it looked like there was any sort of spread. There never was any except for the first altitude update. During the descent, while you were looking out the window, the computer operated just exactly like it does in the simulator. We had blinking altitude and velocity lights from time to time, but this didn't appear to have any effect on the measure of the altitude above the ground. Sometimes it would jump from an altitude of 50 feet to an altitude of 70 feet while we were still descending. At the very end, the indicated altitude agreed with the actual altitude.

10.0 LUNAR SURFACE

10.1 Postlanding Power down

CONRAD — The postlanding power-down went per the checklist. Al called out my portion and his own portion. As soon as we landed, I was spring loaded and had the descent REG-1 OFF and the MASTER ARM ON and blew the vents. I think we were venting within 30 or 45 seconds after touchdown. It went as advertised. We got a descent red light right away.

10.2 Venting

CONRAD — It vented until the ground told us to secure the vents. They told us to secure the fuel vent, and we left the oxidizer open a while longer.

AS12-48-7026

10.3 SITE LOCATION

CONRAD — I was positive of where I was. The thing that confused me was that we were so close to the Surveyor crater. I didn't realize we were as close to it as we were. The thing that confused both of us in the beginning was the fact that distant objects looked much closer. It took us a while to realize that we were seeing many more of the craters that were on our map. They also looked smaller to us on the lunar surface than they did in looking at them on the map. We had a hard time convincing ourselves that the crater in front of us was Head crater.

BEAN — That's right. When you're sitting on the ground, none of the shadows that are visible from the air are visible - the ones that are down on the bottom of the crater - so you end up always seeing the bright part of the landscape, and it's difficult to find the craters. You look out and say, "There's a crater over there." It's difficult for the other guy to see it for a while until you learn to look for the edge. Our plan was to land, discuss where we were for a few minutes, and then make some out-the-window evaluations, I would recommend on the next trip that you make a quick evaluation, knowing that you may not be precisely right. Then make a quick judgment of the general geological features out the window. Don't spend more than 5 minutes at the most on it, because the minute you get out, all these guesses that you were making through the window will be either right or wrong. You can walk behind the LM and look back like Pete did and find that you're standing right next to the Surveyor crater. I think we spent 20 minutes here that we could have better used getting out and getting to work. Then maybe when we got between the two EVA's when we knew exactly

AS12-48-7027

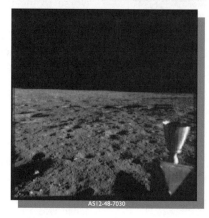
AS12-48-7030

where we were, we could give a better geological description if that's what were in order.

CONRAD — We knew where we were within 1500 feet or closer at the time we landed. We were trying to pinpoint precisely where we were to within 10 or 20 feet, which was ridiculous.

10.4 Light/Shadow Contrast Effects

CONRAD — We have mentioned the light and the shadow contrasts. Whenever we looked cross-Sun, we had the brown effect and we could see shadows. Looking down-Sun, the Sun angle was so low we never could see any shadows except our own shadow from the LM.

10.5 STAR, EARTH VISIBILITY

BEAN — Star, Earth visibility was interesting. We could always see stars at the upper rendezvous window. We could see Dick go by us also.

CONRAD — Al took a quick look around through the ACT and, except for the detents that had the Sun in them, we had lots of stars and no big problem night-adapting to see stars in the ACT.

BEAN — I guess Apollo 11 had a different set of circumstances between the Sun and the Earth clobbering up most of their ACT detents. The Earth was above us and behind. The Sun was low and behind, so our front three detents were in excellent shape.

10.6 Horizon, Sighting Appearance

BEAN — In appearances, it took us a long time to convince ourselves that some of the craters which looked so close were really much farther away. Once we realized, we had ourselves pinpointed and all the craters that we could see.

10.7 PANORAMIC IMPRESSIONS

BEAN — Another thing that threw us a little bit was that preflight we saw some charts that showed little X's where boulders bigger than 1 meter were located, and we didn't see those boulders. We didn't see a lot of boulders lying around on the ground. Looking down in the craters, we could see a few. I think there will have to be some sort of reevaluation of what they're showing as boulders on these Surveyor charts because we didn't see a lot of them just lying around on the ground.

AS12-48-7043

CONRAD — They had me convinced that there were 1-meter or bigger boulders lying all over, and it turned out to be untrue. I'm sure Al's right in his estimate that there were many little 1-, 2-, 3-, or 4-foot secondary impact craters all over the surface. I'm sure that most of these were being interpreted as boulders.

10.8 PGNS Drift

CONRAD — The PGNS drifts were very small. We had no difficulty with our P57's, which we did twice. Our first one apparently was the best. They liked the RLS, and we incorporated our own RLS.

10.9 AGS

CONRAD — The AGS alignment went in an excellent manner, just as advertised.

BEAN — They did.

10.10 AUTO Optics

BEAN — Let me make one comment about the ACT alignments. The stars that we had to use were near the extreme edge of the ACT. When a star gets out there, you can't center your eye in the opening of the eyepiece and view those stars and keep them and the cursor in focus. If you have a choice, try to pick stars that will appear right near the center of the ACT. I think by doing this you can center your eye very well in the eyepiece and you can come up with pretty small star-angle differences.

CONRAD — You can do this in advance, and we ought to use the other 400 stars and unit vectors if that's the case. It's a lot better than trying to use an Apollo star, as Al says, and to find one that's somewhat closer to the center of the ACT. I think it will improve the alignments, and it's no problem loading unit vectors or doing that portion of the program.

10.11 PREPARATION FOR EGRESS

CONRAD — This is one place that we made a mistake. We were a little bit behind and we started to hustle a little bit faster than we should have to get out. We made several mistakes because I allowed us to get off the checklist a little bit. That cost us another 10 or 15 minutes figuring out goofs that we made by simply not staying with the checklist. The checklist covered all items. My hat's off to Scott Millican and everybody else that had anything to do with any of our checklists. We didn't find any mistakes in the checklist. The checklists with respect to the command module and the LM were excellent.

10.12 Evaluation Of Work And Thermal Load In Egress Preparation

CONRAD — The work and thermal load in preparation for egress were very low and they were not fatiguing. We got off the checklist in two places. One place it goes through a very detailed explanation of how to hook up the PLSS for the LMP. It has a statement to do the same thing for the CDR.

69-H-1586

10.13 PLSS and OPS Preparation Donning and Operation

CONRAD — We forgot to hook up my hoses and, when it came time to turn on my PLSS fan, we got into trouble thinking my fan was either clobbered or out. It turned out my hoses weren't hooked up. That was a straight goof on our part. The next goof on our part was when we hooked up our RCU's. There was one switch check that we didn't make. The main switch was in the OFF position and not in the main position. We thought something was wrong with our COMM. Both our switches were in the wrong place, and neither one of us had COMM.

BEAN — Another thing that brought this about is that the gear used down at the Cape have the COMM switches on them but you don't have to use them at all for COMM. The COMM is controlled by the simplex on the back of the OPS on top of those practice PLSS's. Here's an example of the gear we're using not being configured precisely like the gear we use in practice, and that cost us 5 to 10 minutes. That's going to come up again when we start playing with this TV camera, too.

CONRAD — Yes, that's right. There are only two places we got into any trouble, and one was getting off the checklist. The couple of occasions that we got off it, we ran into trouble. The other place we ran into trouble was when we didn't have the gear available. We either didn't have the gear available in the proper configuration or the gear wasn't available. The TV camera was not made available to us, and I'm afraid that's what cost us the TV camera. We were not familiar with it and we'll point out what happened later. Those are the two places we got into trouble. They are places I knew we'd get into trouble, and I'll bite the bullet on the first one. It was my fault for getting off the checklist and hustling, although all it did was cost us some time. We'll both bite the other one on the TV camera.

10.14 MSFN, CSM Conference Communications

CONRAD — MSFN CSM conference communications were outstanding. The lunar surface communications both inside and outside the LM on the PLSS's and on the LM itself were just like having Houston outside the door right next to us. They were really good.

BEAN — I always thought that somebody was located in a building about 5 miles away and, if we would just look back over there behind the LM, we would see Jerry Carr and Ed Gibson standing there talking to us. It was beautiful. There's one other thing that should be added to the checklist. When we went over completely on the PLSS's, there was a period of time when you reconfigured the cockpit before you turned on your PLSS O_2. The reason you don't turn on your PLSS O_2 is that you are going to pressurize your suit, What happened was that I suddenly realized that both Al and I were in the grip of a great octopus because both our suits started to suck down around us, and we started hustling on the check list again so we could get to the point where we could turn on the O_2. In retrospect, I think what we needed to do was to put a little note in the checklist that as soon as you get ready to go over on the PLSS system and you completely hook up on it, you cycle your O_2 once and put a half pound or something

above cabin pressure in your suit and then shut your O_2 off. I think we were breathing it down and we were being very careful to go through our circuit breakers and ECS configuration in the LM, and the suits got tighter and tighter and tighter. We finally got to the point where we could pressurize the suits, and we did. That happened on both occasions, and all I think we need is a little change there to cycle the O_2 and to put a positive pressure in the suits.

10.15 DEPRESSURIZATION

AS12-46-6716

CONRAD — The first cabin depressurization was pretty interesting because as soon as Al opened the depressurization valve after our 3.5-psi check, everything in the spacecraft disappeared out the valve. There was much outgassing, which is not unusual. I had seen it in Gemini. All loose particles that happened to be floating around disappeared from the spacecraft; it gives the spacecraft a real flush. The cabin went right on down to 0.2 and then the spacecraft outgassed for about another minute, I guess, and we finally got down to 0.1 and Al peeled the cabin door open. You have to do this or it just stays stuck until you push it open. It doesn't hurt the door or the seal or anything. We got the door opened and brought on the water boilers which came on in just the right amount of time. In 2½ minutes or so, the water boilers were on the line and we were ready to go.

10.18 LM EGRESS - FIRST EVA

CONRAD — It took me a moment to get oriented and Al gave me a GCA and I apparently was rolled to my right slightly on my way out so that the left lower corner of my PLSS tore about a 6-inch rip in the hatch insulation. That was the only problem - I didn't notice it going out except I did fray the insulation a little bit. I got out on the platform okay and I released the lock mechanism on the MESA and the MESA handle was free in its holder. I tried it and it wouldn't come out. I pulled on that thing as hard as I could pull two or three times, jerked it and everything else and I couldn't get it out. I got tired of wrestling with it, so I just reached over, pulled the cable, released the MESA, and down it went. I went down the ladder and the lighting was excellent. I had no trouble seeing where I was.

CONRAD — At that point, it took me about 5 to 10 minutes to acclimate to what was going on. I didn't have any trouble moving around, but I felt a little rocky. It just took me a while to get organized. This feeling was not bad the second time I got out. As soon as I got out the second time, away I went. So, like anything else, there's a slight learning curve which took all of 5 minutes and away we went. The nicest part of the exercise was that everything went according to the checklist as best as I could see. It went exactly the way we practiced it. And we had no trouble with the equipment. I had excellent mobility in one-sixth g. I missed the fact that I couldn't bend over. That's something I knew I was going to face the whole time and it didn't bother me too much.

Now, I guess the biggest note that I'd like to make, and I think Al and I agree on this, is that the side visor, the side blinkers, blinders were excellent. But you also need a top one. We had a low enough Sun angle that, anytime you put your hand up, looked directly up-Sun, and just blocked the Sun out, you could see perfectly up-Sun. It was only when the Sun was shining in the top of the visor that we had difficulty. So, I think we need to modify the visor so that you have a center top shield that you can pull down and blink the Sun out. If you have that, you can turn 360 degrees and see perfectly in any direction. It will also allow you to look in shadows. The only other time you have difficulty seeing in a shadow is when some other object is reflecting sunlight into your visor when you're trying to look in the shadow. Once you're in the shadow, you can see well. This is nothing new; Neil already pointed that out.

BEAN — Let me say that, just as soon as Pete got out, I had to move over to the right window to take some motion pictures; when I did, I pushed the door partially closed and went to work. About this time, I got a low feedwater pressure. We stood around and tried to figure that out for a while and finally I happened to glance down and noticed the door was closed. I realized what had happened. The outgassing of my sublimator had closed the door, with the result that I didn't have a good vacuum inside the cabin anymore. I quickly dove on the floor and threw back the hatch. The minute I did, a lot of ice and snow went out the hatch. Pete commented about it, and it wasn't 30 seconds until my water boiler started operating properly again. I think that's something that you're going to have to be careful about when you're moving around inside there. I hadn't thought about it before the flight.

10.20 Walking
(Traction,
Balance, Distance
and Direction,
Pace and
Stability)

CONRAD — As I said, we listened to Neil's and Buzz's comments and ours are exactly the same. There's no need to go over them, other than just to remind you to lead your direction changes slightly, but you acclimate very rapidly and it's no problem.

BEAN — I never noticed any slippery surfaces such as Neil and Buzz pointed out. The ground never felt slippery at all to me. The c.g. problems they had were the same. I was very careful not to walk backward; because, I noticed a couple of times when I did, I usually stepped in a crater or on uneven ground and it put me off balance. What do you think about the slipperiness?

CONRAD — I didn't notice any slipperiness, but I think the other comment about moving backwards is the fact that you have such a mass on your back. Al commented to me and I noticed, watching him, that you look like you're standing with quite a forward tilt; but all you're doing is putting your c.g. over your feet and your c.g. is quite aft with that PLSS, so you have the tendency to lean at what at first glance looks quite far forward; it's not.

10.21 Best Rest Position

CONRAD — The position described is a very comfortable position. I never got tired. It's just a normal position to rest in. You can stand perfectly still in that position and rest. Anytime you try to go backward you also have the tendency to stand up a little straighter. Did you feel the same way about resting? Did you just stand?

69-H-1618

BEAN — I never remember doing anything but standing there and never seemed to get tired. As you said earlier, you could work 8 hours out there; if you got tired, you could probably stand against something or just stand there, cool off, and press on. At the end of the EVA, I was feeling as good, particularly in my legs, as I was at the start. The only physical thing I noticed on the second EVA was that my hands were more tired than on the first EVA. I would definitely work on the hands a lot more the next time.

CONRAD — I didn't notice that my hands got tired as much as I noticed that they got sore. When you work for 4 hours and use your hands, you have a tendency to press the end of your fingertips into the end of the gloves; although my hands never got stiff or tired, they were quite sore the next day when we started the second EVA. As soon as you got working again, you forget it. It wasn't until we got back into the command module that we noticed that our hands were sore again. But this was because we did almost 8 straight hours of EVA work which we had never really done before; and I think in one g you don't have the tendency to thrust your hands as far down to the bottom of the gloves as you do in one-sixth g. You really ought to hang onto something up there. It's not as apparent to you when you're working up there that you are pressing your fingers as far out in the gloves, and I think that was just a point.

69-H-1619

BEAN — Another thing that Pete mentioned is that it only takes you about 5 minutes to learn how to move around; the second time you go out, you don't really need the 5 minutes. Neil pointed out that this was the best thing to allow for acclimation. I concur 100 percent. Another good thing is both those POGO's. The mobile POGO that FCSD has is good except that it needs a Z-axis freedom that it doesn't have now. The one on the centrifuge is excellent, I found that running around on the lunar surface, moving from side to side, hopping, and so on were almost precisely like using the one in the centrifuge. I'd recommend having a couple of exercises over there before you go and also recommend changing the terrain over there so that the simulations include a few more big craters, little hills and dales. I think it would be very good training.

CONRAD — And there's another thing - there's no such thing as walking on the lunar surface. Wherever you go, you just want to go at a lope. If you walk, it takes more energy to move slowly and take a normal step then it does to lope.

BEAN — It's interesting and I know we commented about it when we were doing it. If you look at somebody's footprints on the moon, it's almost exactly the opposite of the way they are on the Earth. On the Moon, you can see a flat footprint as the guy lands and then he pushes off with his toe so it ends up being sort of dug in at the toe and flat in the rest of the print. On Earth, a fellow steps forward, lands on his heel, which digs in, and he kind of drives off on his toe. This sort of bouncing along, using your toes for springing and moving from side to side so that the c.g. is always over the foot that's landing, allows you to move out at a pretty good pace and to move a good distance. I had the feeling that, if our TV had been working and if the TV hadn't been pointed in exactly the right place when we went out to 450 feet to lay out the ALSEP, it wouldn't have taken us over 2 minutes to run back, position the TV exactly right, and return to the ALSEP.

BEAN — It would have been no trouble and would have been the thing to do.

10.23 Contingency Sample Operations

CONRAD — We've already commented on the COMM check and voice quality, so I'll go to the contingency sample. It worked very well. The only comment I have is that I probably filled the bag too full. I got some small rocks and then mostly dust and dirt and rolled the bag up later and packed it in the ETB, stashed it on the landing gear like we had practiced at the Cape.

10.24 Transfer of Expendables to Ascent Stage

CONRAD — Then I went ahead and got the batteries and the LiOH canisters with no difficulty. I packed the ETB and our means of transfer worked very well. It took very little time to get the equipment up. The only problem I had was looking directly into the Sun. We were yawed off enough and the Sun was offset enough that I had to walk past, rather than stand directly in front of the hatch. I had to get off to my right facing the hatch, which meant that I had to walk a long ways from the MESA and get into shadows, so that when I looked up I could see the transfer bag and help it over the porch and over the lip of the hatch. I think Al had no problem. I had no problem. He didn't tend to pull me over or anything like we suspected.

BEAN — No. That was good. The only thing that I noticed was that, on the rig that they gave us, there's a small metal pin in the strap that keeps it from accidentally sliding out to the hook and this small metal pin wasn't big enough to prevent this. I just happened to glance down one time and the strap had fallen on the floor. It was just about to go out the front hatch, which would have put a pretty good glitch in retrieval operations. That pin has to be modified. I definitely think that we don't want to go back to a continuous strap. I think the single strap is a workable thing. When you're moving a light load, most of the time, with the exception of the rock boxes, you don't have to use the strap feature over the top of the hook. You can just lean down near the hatch and pull the load in with your hands. It's a lot quicker. When you start carrying a heavy load like the rock boxes, you probably need to use the hook arrangement also.

CONRAD — Our two rock boxes weighed out, Earth weight, at 44 pounds and 52 pounds, if I remember correctly. Neither of those boxes, which obviously were the heaviest things we sent up, presented any problem. The only problem was one that was already mentioned. We knew it was going to happen anyway and I really don't see a heck of a lot you can do about it. This problem is that the lower end of the strap got completely covered with dust and I got dust all over my hands and over my suit arms from handling that strap. I really don't see anything you can do about it.

69-H-1588

BEAN — One other comment. Both times I egressed the LM and tried to close the hatch, it took 45 seconds or so to find something on the hatch I could pull. I think it would be worth the effort to put some sort of hook or something as a permanent fixture on the outside of the hatch so that, when the last man gets out, he can pull the hatch closed without having to grab one of the protective doors over either the handle or the vent valve. It would save 45 seconds or so each time.

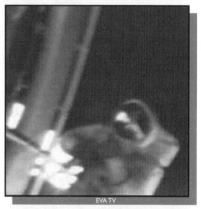
EVA TV

10.25 Egress Observation of LMP

CONRAD — I observed the LMP on his egress. I gave him a little GCA which I don't really think he needed and got the photography of his descent. I started the first part of the TV deployment as we planned. I had the first two pins pulled and the upper door opened at the time Al decided to get out and I had also put up the tripod; I let Al finish the job. You can comment on your adaptability, if you want.

EVA TV

BEAN — I kind of agreed with Pete in the stated area. It just takes about 5 minutes to get used to walking around, and this time should be allowed on the first EVA; once you learn that, you can start easing over and doing your job. I noticed no effect of our movements; I leaned forward and backwards about the same as I expected. I got used to it very rapidly.

10.27 TV Deployment and Operation

BEAN — The next thing I did was get the TV. I think this is where we really got into our first problem. I took the TV off the MESA pretty readily and stuck it on top of the tripod and moved the tripod and the TV over to the deployment place, which was in front of the Commander's window which would be about 10 o'clock at about 20 or 30 feet. The only problem was that, when I got over there, I realized that, because the LM had landed in about a 10- to 15-degree right yaw, the MESA was now in the Sun and that, to put the camera where it could view the MESA, I would be looking directly into the sunlight. If I put the camera over in the shadow of the LM as we planned to do originally, when the MESA was in the shadow, you wouldn't be able to see the MESA. So, I said, "Well, I think I'll take it over and put it on the opposite side, over about 2 or 3 o'clock." I carried the camera over to the opposite side, stuck it there, pointed it at the LM, and called the

ground. It looked to me like there were some pretty bad reflections off the LM and I was concerned that maybe they'd bother the TV. Apparently, that's just exactly what happened; these reflections were far too bright for the TV to handle and it burned out. At least, that's my guess as to what occurred.

AS12-46-6729

Now, it seems to me, this was brought about by two main causes. One, I had personally always felt that we were just carrying the TV along to stand it around and show what we were doing; hence, I personally had never done a lot of serious thinking about how to operate it, the backup modes, its specific limitations, and that sort of thing. As a result, when we got in an off-nominal situation, I didn't really have any good plan for it. I didn't think that the TV was going to burn up from pointing at that descent stage, but I guess I should have been aware of the possibility before I went. I guess that brings up a point. You don't want to make a move with any equipment on one of these flights, even if you think it's not a particularly significant piece of equipment, if you don't understand 100 percent of its capabilities and limitations. Another point about the TV, which I thought about later on, in looking at our plan for how we were going to use it; namely, we were going to take the TV around in back and position it at approximately 4 o'clock so that we could view the offloading of the ALSEP. Also, we were going to use the TV for some 360-degree pans. It's my impression now that either operation would have burned out the TV or given it a pretty good shock.

I think this is brought about by the same situation. I never really thoroughly understood the limitations of the TV. I think that the way we can help a situation such as that, in addition to doing a lot more preflight thinking about it, is to get a TV to work with that's like our flight TV. We need to work with it outside in the Sun using the monitor. If we had done this, I think it would have become very obvious that the TV doesn't have to be in the Sun too long or even point at a bright object too long before the tube is going to saturate and you're going to run into a lot of trouble.

69-H-1585

<u>10.28
Deployment of
SWC</u>

BEAN — The next thing we did was deploy the solar wind collector. That was pretty straightforward. I moved out a good distance from the LM, unrolled it, deployed it, stuck it in the ground. It went in about, I guess, about 10 inches fairly readily and tilted back; it seemed to hold its position very well. Then I started trotting back to the LM. I looked back at it. We were caught in the same predicament of not being able to estimate distances. It didn't look like I had gone out 60 feet, so I walked out, picked it up, carried it out another 20 or 30 feet, and stuck it in the ground quite quickly. Then I stood there, looked at it, and said again, "Well it looked like 60 feet, but now it doesn't." So I pulled it out of the ground again, went another 20 feet, and stuck it in. I probably got the thing out 200 feet, but we wanted to make sure that we got it far enough away so it wouldn't be affected by any of the LM outgassing or anything like that. The final time I inserted it, I pushed down with all the force I could get and put it in about 12 inches.

One thing that continually disturbed us the whole time, particularly Pete, was the fact that the TV cable was right in front of the MESA. Our TV cable laid flat on the ground. It didn't tend to curl up or anything like that; but, because it rests on top of the dust and your feet go beneath the dust, you end up pushing the cable around quite a bit. I think this is a completely unsatisfactory situation and I would recommend that that connector for the TV be moved either over to quadrant 3 or quadrant 1 so that the TV cable would never have to be in the vicinity of the MESA or the area near the front of the ladder. It's just too highly traveled an area to have something like that TV cable underfoot. We never fell over it but it was just a constant problem trying to avoid it.

AS12-47-6898

CONRAD — Yes, I bet you'd be able to use it so that you'd be able to move the cable and get another 40 feet out of it by being able to go plug it in the back of the spacecraft.

10.29 Deployment of Erectable S-Band Antenna

CONRAD — I'd like to talk about the S-band antenna. The antenna was no problem to deploy from the descent stage. I took the antenna right around to the agreed position, which was at the plus-Y gear, and erected it. It went just as advertised, except for the fact that it was not very stable. It was very easy to tip over. When I finally got the antenna completely erected, I went around and got the cable and plugged it in; and even when we had anticipated it (I thought I had the antenna right next to the spacecraft), as it turned out, I just barely had enough cable to connect it. The antenna was exactly in the right place at maximum distance. At that point, we had anticipated that it was going to be very difficult to try to align the antenna with the Earth. One, the sight doesn't allow any latitude. If the Earth is in the sight, the antenna is perfectly aligned. If it's not in the sight, you don't know where it is. So, Al came over and helped because the antenna had a tendency to tip over, especially when moving the crank which moved it in azimuth and elevation. The crank itself tended to tip the antenna over. The crank was stiff. Al grabbed it, pressed it into the lunar surface, and held it as steady as he could

AS12-47-6987

while standing behind it and giving me sort of a GCA on the hand crank. Finally, I picked up a corner of the Earth in the mirror and he still hung on to the antenna while I fine-tuned it. Because we anticipated this problem, it didn't cost us any time.

10.30 LM INSPECTION

BEAN — I don't have a lot to say. I looked at the LM and took the photographs and it's difficult to ascertain how much the high parts of the struts were compressed. I guessed

2 to 3 inches. I saw no skirt damage. I saw no thermal insulation damage. The outside of the LM, as far as I could see, was in perfect condition.

CONRAD — Now the one comment that I made in flight was that there was a rock about 4 by 3 by 2 inches lying right under the engine ball. It hadn't been blown away. I can't figure how it was lying right out at the skirt edge. We took a photograph of it. I don't know whether it will show or not, but it didn't blow away. I was quite surprised after seeing all that dust and stuff flying on landing that it did not blow a rock that size away. We went around and did the PAN photograph and I made a mistake there. I got in a hurry, got off the checklist, and I took all my PAN's at 15 feet. I had Al pick them up later so that we wouldn't lose on the time line.

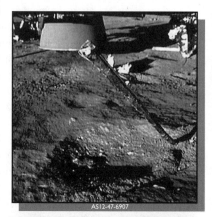

AS12-47-6907

10.31 ALSEP DEPLOYMENT (Traverse, Site Selection and Activation)

CONRAD — We got back to the ALSEP and started a normal deployment. The first thing we noted was that, as soon as we put the packages down on the surface, they began to accumulate dust. Everything went as advertised until Al screwed the cask removal tool on the cask and the cask would not budge. We got the normal fix-it, the hammer; while I beat the blazes out of the side of the container, Al managed to start the cask out. He'd get a little notch on it every time I'd fumble a container. And I really, I guess, started cracking the container. We finally got the element out and the generator fueled. I guess you want to discuss carrying it.

AS12-47-6914

BEAN — Yes, it looked to me like the part that was sticking was about the first inch or so because Pete would beat on it, and it would move out one-eighth inch or so until about an inch of it extended from the cask. Once the element was about an inch out, it suddenly came free and came all the way out, if that will be any aid to whoever is designing the equipment. Some thing was holding it that first inch. After we mated the ALSEP and got ready to carry it out, the workload carrying it out was just about the same as I had guessed from working on Earth. The hard part is holding that weight in your hands.

69-H-1779

BEAN — Even though it is not much, the combination of the weight, the fact that you're moving along, and the fact that your gloves don't want to stay closed tends to make it a fairly difficult task. I would say that it would be acceptable to carry it this way for distances up to 500 feet; but, at distances greater than that, I don't think you want a hand carrier arrangement. You will want to have a strap that fits over your shoulder, or something like that. It's not your legs that get tired; it's a combination of your hands and arms and it just makes you tired. Another thing that occurred that we hadn't seen on Earth is, that as you bounce along at one-sixth g, the RTG package tends to rotate. The c.g. is not exactly lined up beneath the crossbar. It tended to rotate and unlock the two-piece crossbar. This was disturbing because I would have to stop once in a while and relock the crossbar. I would recommend that we definitely put some sort of snap lock on that crosspiece so that when you put it in position and it rotates to the carry position, it locks there. If it had opened up as we carried it, we might have dropped the gear and broken some of it on the lunar surface. It's funny that never showed up in any of our one-sixth g work in the airplane or anywhere else, but it was a continual problem on the lunar surface.

BEAN — When we were at the ALSEP site, it looked as if we were about 450 feet west and about 50 feet north of the position of the LM. It was a pretty good level site. Later when I got back to the LM and looked back, I noticed it didn't look as if the site were that far away. This was the continual problem we had, trying to judge distances. In any event, we had no trouble placing the central station down. We had no trouble putting down the RTG. I did notice, however, that you could feel the heat radiating from the RTG. When I removed the bracket that carried the power cable that ran from the RTG to the central station, it felt warm to the touch. I didn't want to keep my fingers there too long, so I handled it with the ALSEP tool as opposed to just my gloved hands, as I had been doing in practice. That may be something you should practice. Apparently that bracket can get pretty hot, although we only had the element in it a short time.

69-H-1740

CONRAD — I guess the point is, when you fuel that generator, you had better get on the road and get going to wherever you are going to take it. You should get those parts off the fuel element as soon as possible, because they heat at quite a high rate.

69-H-1739

BEAN — The first experiment I put out was the passive seismic. It had two anomalies that I know. One was the skirt. The aluminum foil, the skirt, didn't want to lie down. It wasn't that it had a memory. When I placed it near the ground, the many layers seemed to separate. The skirt seemed to have some kind of static charge on it that would not allow it to touch the ground. It took quite a little pushing to get it to lie down on the ground. The only way I could make it lie flat was to put a little dirt on it, which I tried. But that wasn't a very good idea because it's difficult to put

69-H-1736

little clods of dirt on it. I later got some Boyd bolts and made little alignment tubes to sit on it. That worked real well; it held down the skirt pretty well. The second anomaly was that the little dish the passive seismic sits in needs to have a solid bottom so when it is placed on the ground, there is no danger of dirt easing up through the center of it, as there was in the case of our dish, and touching the bottom of the passive seismic itself and causing a thermal short that would ruin it. We spent quite a bit of time tapping out a nice neat hole so this wouldn't occur. Really, I think the fix should be to put a solid bottom to that dish. I think the addition of the bubble level to the top of the passive seismic was a good one. I noticed it was really easy to level the experiment with that. While I was doing it, I kept an eye on the little BB in the bowl-leveling scheme. It was just rolling all over the place as it had for Buzz when he tried it on Apollo 11. I don't think that's the way to go for any other leveling. I think this bubble works real well, and it works pretty fast. I'll go ahead and cover the others that I laid out.

The magnetometer was a beautiful experiment and it was easy to deploy. It was easy to align and level and it took quite a bit less time than the passive seismic because it was sort of self-contained. You just screwed the legs to make it level. You could grab one of the magnetometer arms and move it around so that it would be in alignment. In both cases, we aligned them exactly zero, zero, or exactly East.

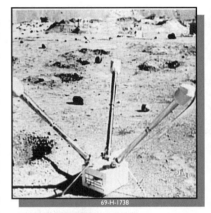

The suprathermal ion detector and the cold cathode ion gauge, which were combined experiments, were difficult to align and we knew this before we left. The legs are too close together for the height and weight of the experiment. When you try to get it on the ground, it just wants to tip over. The little place where you insert the ALSEP tool is on the end that has one leg; it's a three-legged configuration, and so that tends to, if you put any offset or force on that attachment, tip it over. Next, the cold cathode ion gauge. A screen comes out of the side of the container in the suprathermal ion detector. It is spring-loaded so that it will hold a flat position when it goes over center. It's a real problem to keep the side of the suprathermal ion detector balanced while you try to make this screen go over center so that it'll lie flat. It took a couple of minutes to get the screen to lie flat. These spring-loaded devices are a real pain up there. You should have one that doesn't have any spring load to it. You open it up and drop it on the ground and it just lies flat from its normal weight instead of having some spring-loaded, over center device. It just adds time and work trying to get these little devices to work properly. The cold cathode ion gauge comes out of the side of the suprathermal ion detector. The cable was so stiff that if you put the gauge in the proper position, the gauge itself was so light and the forces on the cable were so strong, it would just pick up the

gauge and move it to a non-desirable position. It ended up with both of us, Pete and I, working together. He held the ion detector while I tried all sorts of different deployment angles of the ion gauge to finally get one that would work. The only way we could make it work after spending about 5 minutes on a 30-second job (if it had been designed properly) was to lay the ion gauge on its back so that the front end pointed straight up at a distance that wasn't near as long as the cable itself. This seemed to work pretty well.

When we were deploying the suprathermal ion detector, the lid came open a couple of times. This lid was supposed to be deployed from ground command after we had left the area so that the exposed mirrored surface would be nice and clean and the two detectors would not get dust in them. I'm pretty sure that we did get some dust on the top of it. I hope it's not enough to bother the operation. I've got two other comments. My experience working the Boyd bolts is that you can do them a heck of a lot faster if they don't have those little alignment tubes on them. I don't know how Pete feels about this, but I recommend that you throw those off and just use the Boyd bolts. I could always stick my tool in there a lot faster when there were no tubes, and I can also see when the little bolts jump up a lot better. The second comment; I think we're kidding ourselves if we think there is any way to deploy this experiment without getting a lot of dirt and dust on it. The pictures are going to show this. They just have to be designed to accept dirt and dust. If they can't accept the dirt and dust, then they are going to have to be packaged in some way so they can be deployed completely and then, the last act would be to pull some sort of pin and flip off the covering that would have all the dirt and dust on it, exposing the nice clean experiment.

CONRAD — The only one I deployed was the solar wind spectrometer and it went exactly as advertised. I checked the four legs down, took it out the proper distance, aligned it, and turned her loose. The Boyd bolts, as Al pointed out, were no problem; it would probably be easier if the cups were lower. The bolts should be kept covered with the tape though, because of the dust problem.

10.32 Sunshield Deployment

CONRAD — The sunshield deployment worked perfectly well and as advertised in one-sixth g; it popped up, lifted off the ground, actually. It was a real thrill. The antenna alignment went as advertised. I had played with it enough that I knew how to align it correctly. Apparently it is aligned all right, because you are receiving good signals. We feel we did most of our homework on ALSEP.

10.33 Visibility of Boyd Bolts, Light Piping with UHT, Decal/Label Legibility

BEAN — I do too. I think we knew exactly what to do on it. One thing that turned out a little different than I imagined; I had difficulty reading the decals that we had put on the ALSEP. As I recall, looking at it on the surface, it looked as if there were black writing on silver background. When I tried to read this on the lunar surface, it was very difficult. The brilliant light reflected off the silver and you couldn't see the black. We had the sequence of laying that ALSEP down pretty good, and I would recommend on the next one, they use a black on orange or something like that to decrease the amount of reflective light off the decals. It's going to be needed.

10.34 Selected Sample Collection

CONRAD — The selected sample collection, I felt, took a fair amount of our time. We did not collect quite as many rocks as I would have liked to on the selected sample. However, we did go over to the large crater, collect some rocks from there, photograph them, and then return to the LM. We were beginning to run out of time at that point, although the EVA had been extended and we did get enough rocks to fill the whole rock box.

10.35 Sample Return Container/Core Tubes

CONRAD — The sample return container worked as advertised. I do want to comment here that I feel the practice we did on the K-bird was excellent for one-sixth-g work. Opening and closing those boxes on the K-bird took away all the surprises. I was ready for some of the heavier forces, and had a handle on them. I was aware of the fact that the cable hold down was going to tend to rise when they were holding the boxes down. All the work we did on the K-bird was excellent, and made things a lot easier on the lunar surface with respect to the rock boxes. Al drove the core tube.

BEAN — The core tube was pretty easy to drive. I think we must have been in a different sort of soil than Buzz was in; I augered it a bit as I drove it in, but I really never had the feeling that was necessary. I think all that was necessary was to hit it pretty doggoned hard and drive it in there. We had no trouble going the full length. A couple of things I noticed as we worked was that whenever I held onto metal tools for any length of time, anything shiny like the extension handle; the tongs; or later during the second EVA, when I was carrying the hand tool carrier; that my hands would get warm. If I would put them down and remove my hands from them, my hands would get cool again. It was not too hot to handle; it was just the fact that I would notice they started to warm up.

Another general impression I had working with all the equipment is that the lunar equipment we have is generally too flimsy. If we are going to work with this gear, we should beef it up so that we don't have to be so careful about breaking it. I was always concerned that I might actually break some of the ALSEP equipment. I really don't think we need to be quite that tender with it. The

same way with some of the tools we used later on. Pete may not agree with it, but he probably should say something. I got the feeling when we were working on those selected samples and coming back, we needed two things. We needed a little bit bigger set of tongs so that we could grab bigger rocks. We always ended up being able to get little bitty rocks. If we had a bigger set, we could get the little ones, and also the big ones. With that little one, you just end up getting little rocks. You want to reach down, get down, and look at rocks, and you want to pick up things. You don't always want to stop and use those tongs. I had the feeling that if we just had a strap mounted on our back, or to one side of our back or something, we could work as a team. One fellow could hold the other while he leans over and picks up or inspects a rock, or looks in a hole, or whatever he wants to do, and then lift him back up and he wouldn't get dirty.

We had talked earlier about just falling over on our faces and catching ourselves on our hands, or getting down on our knees, and inspecting whatever rocks we wanted to look at. When we got there, we could have done this physically, but the problem was, it was just so dirty that you didn't want to do it. I went down on my hands a couple of times, but each time I did, I went down where I would land with my hands on a rock. I would stand there until I saw what I wanted to see, and then do a kind of push-up from the rock. But there isn't always a rock around to do this sort of thing. If we just had some simple strap, worked as a team and got the big rocks fast, and looked at what you wanted to real fast, I don't think it would interfere with anything else you did.

AS12-48-7149

CONRAD — I agree with all that Al said. There's no doubt about it, you need a bigger set of tongs. By the same token, you need a bigger set of sample bags. Those big Teflon bags are very unruly in the lunar environment. They appeared to get brittle. They took a set and were hard to straighten out. They just didn't handle at all well. I think we need to develop a cloth bag of some type that will maintain its shape and not take a permanent set. These bags had been folded and tended to take a set, and when we straightened them out, they tended to be brittle. They had several long cracks in them when we wrapped up the rocks. One other item in the first EVA, the colored chart, I took out because I could not bend over, and there was no reasonable way to stick it in the ground. I tried to work it into the ground so that it was perpendicular to the Sun. It didn't work because of the soft dirt. It fell over and became covered with dust. I got it back up and tried to brush it off, but it was impossible. I just made a complete shambles out of it. The dust clung to it so badly that we didn't get a color shot of that.

10.36 LM INGRESS - 1st EVA

CONRAD — Neither of us had any trouble ingressing the LM and getting the door closed. The cabin repressed right away, and we went right into the PLSS recharge.

10.37 PLSS Recharge

CONRAD — We had practiced the PLSS recharge several times, and it paid off. The PLSS recharge went as advertised. The equipment was easy to handle at one-sixth g. All the stowage was adequate. We followed our procedures to the letter and we never fell over any equipment. It all

AS12-47-6932

went in the right places and transferred around as we had practiced. I wanted to weigh the water. I put in this 25-cent scale, which should be set at zero. If anybody had thought about it, including myself, the spring tension in the scale itself was never zero in one-sixth g. As I unscrewed it to zero, I unscrewed it all the way; and the screw, the spring, and everything disappeared into the bottom of the scale. We had a slight amount of difficulty, like 25 minutes, putting that baby back together again, which we finally did. It would be wise to put in a reasonable scale that can be zeroed in one-sixth g. Let somebody think about it a little bit, we can plan this one better. The water weighing went okay. We both had plenty of feedwater left. The levels passed to us indicated that we were working 900 to 1000 Btu's. Both of our oxygen supplies recharged to more than 80 percent. I don't even know what the top-off was; I didn't even notice.

10.38 LM EGRESS - 2ND EVA

CONRAD — The second EVA prep went a lot smoother. We stayed with the checklist. We had gotten over our excitement and we did it in a much more orderly manner.

10.39 Geological Traverse

CONRAD — We covered a lot of distance on this EVA. We were told that we went more than a mile. We used our chart while on the surface. It was a good thing we had it along. We stopped and consulted it on several occasions. Although navigation is not difficult out there, you really do have to stop and pin down certain craters. They wanted us to go to specific craters, and I think we hit all of them except Halo crater. I knew we were in the general area. Several craters were there, and the photograph map was not clear as to which crater was Halo. We may have sampled the wrong crater. If so, it was very close to Halo crater and should have accomplished essentially what they wanted us to pick up with the double core tube.

69-H-1589

10.41 Contrast Charts

BEAN — I want to discuss number 41 which is contrast charts. We had three of them. One of them got dropped in the dirt and was completely covered with dust; so it was useless. There was no way to dust if off.

AS12-48-7036

CONRAD — We hung them on the corner of the table. We had three of them hanging on the corner of the table. When I removed the first LiOH box on the first EVA to send it up, that one fell off and I had to pick it up out of the dirt. Once it gets in the dirt, forget it.

BEAN — There's no way to dust anything off there, which brings up a good point that we'll be covering in a few minutes concerning when we got back in. I took one chart and put it on the sunny side (and you can see all the different grades of gray), took pictures of it, and reported that over the air. I put the other on the shadow side, and the only thing that keeps you from looking into the shadows is if the Sun is low

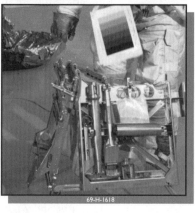
69-H-1618

enough so that you can't get your hand up and shield your eyes from the direct ray of the Sun. If you can do that, you can see down in the shadows just like you can on Earth. I was able to do this in this case and so I could see the full range of blacks in the shadow. Pete's earlier idea of putting an additional opaque visor on the top of the helmet is a whale of a good idea. You could adjust that thing for the sun angle; you could wander around looking up-Sun, down-Sun, and across-Sun; and it wouldn't bother you a bit. I think that's one of the best suggestions that we got right there.

**10.42 Samples
(Collection and
Photography)**

CONRAD — Our documented sampling went exactly the way we practiced it on Earth. The sample bags were too small. There were many samples we had to put in that were too big to go in the bag.

BEAN — One of the best things about the charts was that we had (this was Pete's idea during the last month) traverses planned for three different places that we might land. That is what gave us the capability to go out there and do some good geology. They have done a lot of thinking about what to look for. We named some of the craters, we knew the traverses, and we were able to massage one of the preplanned traverses a little bit. We were able to follow the traverse pretty well. It looks like we can land now on one spot; thus, you want to practice the exact traverse on Earth that you are going to be doing on the lunar surface. I might even suggest that they make those little

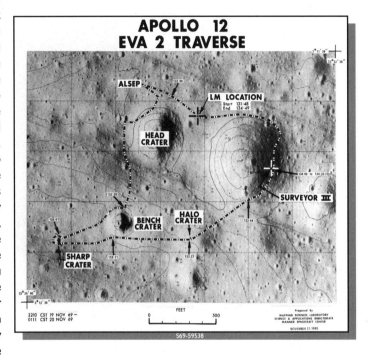

APOLLO 12
EVA 2 TRAVERSE

craters out in Flagstaff, Arizona, like your particular site. Then, you can do the exact traverse during training that you are going to do during the mission. This allows you to save a lot of time by going to the proper side of the crater and trenching and core tubing at the right place. Then on the Moon, you can follow the same preplanned traverse and get a lot more done much faster.

CONRAD — This was the trade-off. I knew it was going to happen. They wanted distance and documented samples on five specific points. Only on occasion between those points did we stop when we saw something that was different and sample it because we had to hustle all the way to cover all that ground and to make those five points. We really had to move out; therefore, there was a certain amount of compromise between the amount of sampling we did at each point and in between these points. There was no point to which they sent us where we couldn't have spent at least an hour very easily. Had we had a little bit more time to experiment, one thing we might have tried (and it crossed my mind at the time, but I thought if we do that we're going to get tied up) was to use the tether and get one of us down in the crater that had the melted looking rocks in the bottom of it. I think that would have been a real boon. But the tether wasn't really long enough, and the crater was pretty steep. I'm not so sure how well we would have gotten back out of it.

BEAN — The entire lunar surface was covered with this mantle of broken up material, fine dust of varying depth. As a result, everything looked pretty much the same, sides of the craters, tops of the craters, flat lands, and ejecta blanket. If you're going to do any geology, you're going to have to dig through this mantle of brown or black and to look beneath the surface a little bit. We had a shovel that we used for trenching, but because of the length of the extension handle and the inability to lean over and what have you, we never could trench more than about 8 inches. That was about the best we could do, and that was a pretty big effort. If we're going to do any good geology, it's going to take a lot of trenching to get down below the surface. I'd like to recommend that we get a better trenching tool. Maybe all we need to do is lengthen the extension handle about 6 inches; but if we're going to look and see what's beneath the surface, we're going to have to dig it out of there somehow. I also recommend that we get a lot more core

tubes aboard the next flight. I felt that, on the surface everything was pretty much the same and the real secrets were hiding about 2 to 8 inches under the surface. We really need to scrape away the upper surface or core down through it.

AS12-48-7090

CONRAD — The environmental and gas samples went as advertised. I did notice on both of them a tendency for the threads to be a little bit sticky, as if there were a vacuum welding or maybe a slight expansion that was causing some drag. It wasn't bad, but it did not need to be much worse before causing difficulty in closing both the gas sample and the environmental samples. But, they worked okay.

10.43
Surveyor Site

CONRAD — When we looked at the Surveyor from directly across the crater, it looked like it was sitting on a vertical wall. It really looked steep. My first thoughts on the first EVA was that we're going to have one whale of a time getting down to it. At that point, it was sitting in the shadow; but by the time we got out on the second EVA, it was no longer in the shadow, It then gave the appearance of not being on so steep a wall. Sure enough, when we walked around the other side of the crater and got to it, it was on about a 12-degree slope. Just angling down the side of the crater to it was no problem whatsoever. No difficulty was encountered working around it. We had been concerned about it sliding down on us. However, it was firmly planted in the side of the crater. The lower gear was well dug into the ground, and there was no chance of it sliding. I actually pushed on it several times; there was no tendency for it to tip over or to move. So we went about our business just the way we practiced it at the Cape.

69-H-1982

69-H-1616

I read the checklist and Al took the photographs. The only trouble was one tube which I couldn't cut. I bent on it as hard as I could and I hardly made a dent on it. I don't know what kind of a metal tube it was but I don't think it was quite up to snuff. It had to be a thicker wall tube than they told us because the five TV tubes were cut with hardly any effort at all. So we got everything but the glass sample. The glass was bonded to the metal sheet. We curved the metal sheet but all we got was fine slivers of glass, so we stopped messing around with it.

BEAN — We did get a tube, however.

CONRAD — Yes, it was a support tube to the large electronics box.

BEAN — We noticed the Surveyor had turned to sort of a tan appearance, including the white parts, the chrome, and the shiny parts. We looked at it closely and rubbed it. You could rub off this brown color if you rubbed hard enough. It gave you the feeling that it wasn't blown on when we flew down in the LM, or rather that it had adhered to it over the years it had been in the crater. We took enough pictures so that we can document this. It wasn't very difficult, I didn't think, to operate on that slope, Pete. It wasn't particularly slippery.

One of the things I wondered about beforehand was, when once we got down the slope we wouldn't have a good sense of the vertical and we'd tend to lose our balance; but that wasn't the case at all. It was just like a 12-degree slope on Earth. I didn't have any tendency to slip down, and I wouldn't hesitate to try a steeper slope. But I do think that having a strap or tether that you could use, in the event that you got down into too steep a slope, to help you get back out is a good idea. It probably ought to be a standard piece of equipment on the following missions. As Pete brought out earlier, it would have been nice to go down to the bottom of that slope to the material that looked melted. Our strap was only about 10 meters long, and I would recommend that, as a standard piece of equipment, that you put a strap about 30 meters long in a saddle bag or somewhere in your equipment. That way, you could help a man down the side of a slope. He could just carry that strap down to the bottom and pick up any rocks he wanted to get, and then you could help him back up. I don't think you would have any trouble; you would have to use discretion in case you got half way down and found out the sides were a little more slippery than you thought. The only thing I noticed about working on the inside of the slope, Pete, was when we tried to walk out of it, it took a lot more work because you couldn't bounce from side to side and spring off your feet like you could on level ground. It just wore you out a little bit more, and I wouldn't be surprised if our heart rate wasn't up a little higher after coming up out of the crater. But there was never any danger of slipping down in the bottom or anything like that.

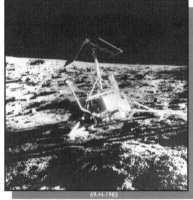

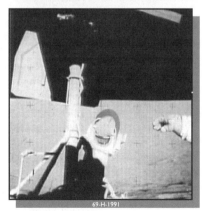

CONRAD — I guess I ought to finish off one thing on the Surveyor. My hat's off to Al and Joe Roberts and all the people that worked on that bag we carried on our back. It worked out extremely well. Al's suggestion made earlier, but not on this debriefing, was that we should consider carrying a lot of equipment on the back of the PLSS for documented sampling and so forth. It's better than carrying that tool carrier around. You could hang a considerable amount of the gear on the back of two PLSS's and you work as a team. I think it would be a lot handier.

BEAN — I do, too. The big pain with that tool carrier is that you have to hold it out from your body so that your legs don't bump into it as you walk, which means you have to hold it by one hand. That's not a big deal when it's light and there are no rocks in it; but when you start filling up with rocks, it gets to be a pretty good stunt to hold it out there for long periods of time. I was running two and a half miles a day towards the end of the training period to get my legs in shape, and my legs never suffered a bit. If I had it to do over again, I would run about a mile a day and spend the rest of the time working on my arms and hands because that's the part that really gets tired in the lunar surface work. If we could somehow eliminate that hand-carried tool kit and mount those things on the back of the PLSS's, half on the CDR's and half on the LMP's, you'd be able to move around the surface a lot better. Your hands would be more free. One could carry the shovel, one could carry the tongs, one could carry the gnomon, and you could end up doing better work faster by that technique than with your hands full of hand-tool carrier.

AS12-48-7133

CONRAD — I had about 20 Earth pounds extra hanging on the back of the PLSS. It caused me no c.g. problems. I ran across the lunar surface, and it was not that tight. The bag and the camera itself were flopping around back there, and they didn't bother my stability one bit. I just kept whistling. Our only concern was that it might fall off or something, which it didn't do; and so I think it's a good idea.

AS12-48-7100

BEAN — I was going to say we probably ought to make a comment here. I know that they are trying some sort of wheel vehicle for the next flight. My impression was that you could use a wheel vehicle but you probably should have one with wide tires.

BEAN — Although the dust was only an inch deep or something like that, if you had some skinny tires it might give you a little problem. I don't know how big in diameter they ought to be, but they ought to be fat things to help it ride along the surface.

69-H-1617

10.44 Sample Stowage

CONRAD — The sample stowage on the second EVA worked out perfectly. I think we had the box packed as tight as we were going to be able to pack it there on the lunar surface. We wound up with four large rocks in a Teflon bag which we brought back for an additional 13 pounds of rocks. I've got one comment here and I know that this is one of the problems with not preplanning something. I know everybody, including ourselves, agreed with a preflight criterion of one EVA extension, but it broke my heart to get back in that LM and find out that we had 6 hours in those PLSS's and Mission Control hustled us back in after 4 hours. We killed 2 hours sitting in the LM - we wasted - we actually sat on our rear ends and did nothing for 2 hours. We weren't tired and it's really a shame that we did not get a second extension on the end of that EVA because we hustled past Block crater and back up to the LM. I was hustling Al because I felt that he had committed us to get in at 4 hours and nobody changed their tune; so we were practically up the ladder, and that's really a shame. I think we've got to be open minded on Apollo 13 and subsequently. If these guys are in good shape - and there's no reason to believe that they won't be (we were in excellent shape) - then let's not hustle when we don't have to. As a matter of fact, we could have gone another REV down there on the lunar surface before lift-off, and it wouldn't have perturbed anything.

69-H-1774

BEAN — I agree 100 percent with Pete. That was a real shame. If there was anything in the whole flight that should have been done differently, we should have gotten in maybe an hour later right then. Like Pete said, we got in and we sat around for a couple of hours there just waiting for the time to start working on the ascent checklist.

AS12-49-7318

10.45 SWC Retrieval

BEAN — We performed the solar wind collection retrieval, and that didn't work like I was hoping it would. When I reeled the collector out, it came out very nicely; and the foil itself seemed pretty flexible. At the beginning of the second day after we woke up, it looked like the foil had taken a set around the pole. It definitely had taken a set of some sort. When I tried to roll it up following the second EVA, it rolled up about 6 inches and then didn't want to roll any more. It wanted to crinkle and tear although I was very careful with it and tried to recycle it a couple of times like a roller shade. Finally, it did tear about a 6-inch longitudinal rip; and I realized then that it just wasn't flexible enough and didn't want to roll up. So I let it go and let it sort of window-shade all the way around and then tried to roll it up by hand and not get my fingers on any of the foil. I'm sure I wasn't able to do this entirely. I expect there is some dust from my gloves on the foil but

AS12-49-7278

I did the best I could. I understand that it is possible to dust that off, bake it off, or something like that and it doesn't bother the experiment.

As a result of this, the rolled-up experiment was larger than the bag it was supposed to fit into; so I had to take and crush it with my hands - kind of squeezed down on the foil. It made it look sort of full, but I don't think it degraded the experiment at all. I would like to recommend that before the next flight goes up somebody take a look at what is actually happening to that foil as it sits out in the lunar environment. It may not be the foil that's presenting the problem. It might be the tape that they are using on it actually cracks or gets stiff or something. It may be the same effect they were seeing with those Teflon bags. They may take a set and it doesn't want to roll up any more. This does compromise the experiment somewhat.

CONRAD — The final comment that I would like to make is that Ed Gibson and Joe Roberts and his people did an outstanding job. Our checklist and procedures were well documented. We were able to handle all the contingencies that came up, and I think most of the success of the operation is due to their careful preflight planning and excellent work that they did on the EVA's.

AS12-57-8448

BEAN — All right. I think also included in there are the geologists Uel Clanton and his group and also USGS Al Chidester and his group. I think one of the best things was this preplanning of the traverses. That really saved us a lot of time and let us get out and go to the places that they thought were geologically interesting and allowed us to do the right things at the right places.

10.46
Close Up
Camera
Operation

BEAN — As a result of having to hustle there, I don't think I really gave a good enough effort on this close up camera. I did as much as the time allowed, but the time wouldn't allow me to go out to those three different types of soils that we had seen and to take pictures. I couldn't go down in some of the craters that had some good glass in them or glass-topped rocks. I couldn't go out and do many of the things that the close up camera is very good for because there just was not time available.

AS12-57-8452

10.47
LM INGRESS -
2nd EVA

CONRAD — The second ingress was the same as the first. No problem; we went right down the checklist jettisoning the equipment. The only comment I have is we shut off our PLSS feedwaters as we did on both ingresses down at the bottom of the ladder. My recommendation is that we take a fixed time prior to ingress, and even an earlier time like 10 or 15 minutes before ingress, and get that PLSS feedwater OFF because, when we dumped our equipment on the

AS12-57-8455

third cabin depress, we got quite a bit of water and ice out of the PLSS sublimators. I think the more you can dry out those boilers before you get in the better off you are. Otherwise, the equipment jettison procedures and everything worked extremely well, and we got rid of all the equipment.

BEAN — One comment about the COMM between MCCH and ourselves up there: I don't know from their point of view what they thought, but we were getting about the right amount of information every time we asked a question; and the answer seemed to come up pretty rapidly. From our point of view, the COMM between ourselves and Ed Gibson and the experimenters was very good. I an hoping that if they had any questions they did get to us and we answered them at the time or, at least, were able to look at what they were interested in so that we know the answers for them right now.

CONRAD — One other comment on the second EVA: we had a problem with the camera. The wheel on the bottom screw that locks the RCU plate and the trigger to the camera is a machine wheel that fits on the end of the screw. It was a press fit, and because of the thermal configuration and the tolerances of that fit, the wheel actually fell out and the camera came completely apart. The trigger came off the RCU guard. That was the end of the usefulness of that camera as far as we were concerned.

BEAN — I would like to say something about that camera. We got a lot of dust on ourselves and also on the outside of the camera. We kept looking at the lens to see if there was any dust on it and to see if it was going to degrade the pictures. Neither Pete nor I could see it on each other's camera, although the other parts of our camera were covered with dust. We'll have to take a look at the pictures that we returned. If it does turn out to be a problem, we're going to have to come up with some sort of brush we can use to dust off the lens, because I don't see any other way. We were trying our best to keep the equipment clean; but just moving around, trenching, leaning over, and all the other things tend to get dust on the equipment.

One other thing, when we go back to the LM, we tried to dust each other off. Usually, it was just Pete trying to dust me off. I would get up on the ladder and he would try to dust me off with his hands, but we didn't have a lot of luck. We should have some sort of whisk broom on the MESA. Before we get back in, we'll dust each other up high. Then the LMP will get on the ladder, and the CDR will give him a dust or vice versa and then will get on in. We are bringing too much dust into the LM. Another possibility is that just as soon as you get in you slip on some sort of second coveralls that fit over the feet on up to the waist, because that's the dirty area. Then you keep that on all the time you're in the LM and take it off just before you get out. The other alternative to this is that you put on a similar something when you're getting out onto the lunar surface. The reason I suggested the former was that I think you want to be as free as you can possibly be when on the lunar surface. Adding another garment over the top of the already existing equipment is going to be restrictive and might give you a few more problems.

CONRAD — I got quite concerned with not only the wear and tear on the suits but the effect of the dust on the suits. On our final hookup back on the LM ECS system for ascent, it was all we could do to get our wrist locks and suit hose locks to work. They obviously were beginning to bog down with dust in them. When you go over these suits later, you'll be able to analyze this. I have no idea what the effects were on the O-rings. Suit integrities did stay good, but there's no doubt in my mind that with a couple more EVA's something would have ground to a halt. In the area where the lunar boots fitted on the suits, we wore through the outer garment and were beginning to wear through the Mylar. I'm sure that with all the wear on the outside surfaces there's bound to have been rubbing of the bladder. I'm sure they will be very carefully inspected to see what these effects were. Al and I had extreme confidence in the suits; therefore, we didn't give a second thought to working our heads off in the suits and banging them around not in an unsafe manner but to do the job in the way we had practiced it on Earth. These suits were more worn than our training suits. We must have had more than a hundred hours suited work with the same equipment, and the wear was not as bad on the training suits as it is on these flight suits in just the 8 hours that we were out. I think it has to be the abrasiveness of the dust.

BEAN — I think that one of the best decisions that we made was not taking the suits off at night. This allowed us to control our temperatures pretty well. We were always able either to turn on the LCG pump and get cool that way or to turn on the vent flow. If we had had the suits opened up, I'm afraid that we would have had a lot more trouble with dust in the zippers, inside the suit, and inside the helmets. It was tough enough on just the wrist rings and neck rings. We tried to wipe them off before we put our equipment back on the next morning but we did notice it harder to put on. I didn't have any leak rate for all the pressure checks prior to launch and at other times, but during the last pressure check that we pulled I had a leak rate of something like two tenths over the minute. So, the thing was leaking somewhere and it must have been around the neck and wrist rings because those were the only openings that had changed.

CONRAD — My suit was the same way. I had about 0.15 over a minute although I had very little on our first check prior to getting out.

10.51 REST AND SLEEP ON LUNAR SURFACE

CONRAD — I made a technical error before I left when the suits were sent back to ILC and the boots were put on. We knew that we had to refit the suits, and I let myself get conned into refitting my suit in long underwear and not with an LCG because the flight LCG was PIAed. That was a mistake. I wound up with the legs being too tight. I realized this prior to lift-off while staying in the suit for a long time. I had spent only about an hour or so before in it fitting it in my long underwear, but it became unbearable that night. It spoiled my rest period. I did not want to take the suit off so I stayed that way all night.

I slept only about 4 hours, and it was mainly because of suit discomfort on my shoulders. The next morning, Al did an outstanding job on letting my legs out for me, which took him about an hour. As far as the remainder of the rest went, the cabin temperature remained good all night. I didn't notice any change in the temperature all night. I didn't hook up my LCG. The only thing that I noticed was that the very bottom part of my legs from the knees down to my feet tended to get hot while I sat in the suit with no air. Although they weren't uncomfortable, I had the feeling that I was beginning to perspire down there and that after a while it was either going to get wet or it was going to get cold or both. So, about every 3 hours, I put my suit hose blue to red and blew my suit out with dry air which took all the moisture out of the lower boots and dried out the LCG socks. I would let it blow for 5 to 10 minutes and then would take it off, and that's the way I remained all night long. I never used the LCG pump. I don't know whether Al did or not. I was never too hot or too cold. The hammocks were excellent. The first 4½ hours I slept, and it was a good sound sleep. The only reason that I couldn't sleep the rest of the night was that my shoulders were so uncomfortable in the suit. My feet were plastered against the bottom of the suit and my shoulders were plastered against the top, and there was no easing it no matter what I did.

BEAN — The only comment about my hammock concerned that Beta cloth covering down at the foot end of the hammock. I wasn't able to tighten the hammock completely. I had to unsnap it and pull it back before I could tighten the hammock. I think that Beta cloth ought to be changed, modified slightly, so that you don't have to dismantle it so much to tighten the hammock. When we got back in, the ground said, "You can put 15 pounds of rocks," I think it was, "behind the OPS's." That's where we did put them and they set very well in there. We used the tie down and tied them in there. It was a good place. Then they said we could put 25 pounds in the left-hand side stowage compartment. We didn't have the 25 pounds there, but I wasn't briefed, prior to lift-off, exactly how we were going to put the 25 pounds in there. Were you, Pete?

CONRAD — No, and I didn't quite understand the comment because both sides of the left-hand storage compartment got thrown away.

BEAN — All I can figure was that somehow we weren't supposed to throw part of it away. We didn't have any rocks in there anyway but, if we had, I guess we were supposed to maintain part of it or something. This is something that we should have been briefed on before we went. If we had an additional 30 minutes or so, I think we could have grabbed 25 pounds and stuck them right in there.

CONRAD — This is one of these typical things, though. About 6 weeks ago, they went to Jim McDivitt and told him that there was no doubt that we could collect more rocks than would fit in the rock box and we needed some clarification. He took immediate action to find out about how much more rocks we could carry. Over that period of 6 weeks, people came and went and hemmed and hawed and changed their minds. How many extra rocks we could carry, and where, was one of the very unclear things when we left. That left-hand side stowage thing came completely out of the dark when we were on the lunar surface. I never heard of it before flight. The only discussions that I had heard were that we couldn't stack on top the OPS's because the OPS leg holders were designed only for the OPS's and that part of the structure wouldn't hold. But anything that we could stash on the floor behind them was okay as long as we kept it off Z27 bulkhead. Then they modified that weight once in flight and they came up with the left-hand side stowage, which didn't make sense to me. I wasn't going to argue with them because we didn't have that many extra rocks, but both left hand side sections of the stowage got thrown out on the lunar surface at one time or another - one at the end of the second EVA and one on the third dump. So either somebody wasn't following our procedures or was not aware of them, and that again is an example of the last minute Mickey Mouse and unplanned things that are either unclear or cause problems. In defense of everything, I think we had very few last minute changes. I think we were in better shape than most flights on last minute changes.

BEAN — I don't know anything to add, Pete. That's exactly right. That was just one of the unusual ones.

CONRAD — We cooled our heels for 2 hours in the LM, ate more food, which was good, and picked up at minus 2 hours and 50 minutes on our lift-off checklist. It went exactly as advertised to the point that a couple of times I called the ground to find out if they were still around. We went through that thing by ourselves, absolutely per checklist, and counted it right down to lift-off. Nothing was done that was not published in the checklist.

BEAN — Two things I noticed. One, after I got the COMM set up, I noticed that it wasn't the same COMM that we had during descent. The difference was that during ascent we had VHF A receiver OFF and during descent we had VHF A receiver ON. I kind of thought it should have been ON for backup, but we asked the ground about it, they looked at the checklist, and said it should be OFF. We left it OFF, but I later turned it on during the rendezvous. That was one little anomaly that was probably our own fault for not catching it during training. The other one was when Pete performed the RCS checkout. When he fired the thrusters on the right side, it knocked over the S-band erectable antenna; so we switched over to the spacecraft S-band which didn't even lose lock. We got some good movies, I think. He fired some of the thrusters, and I took some 16-millimeter movies out the window. Hopefully, the geologist can get some feel for movement of dust with that engine and maybe compare or extrapolate down to the descent engine.

CONRAD — I'm going to have Mission Control look over their data, but that RCS firing on the ground appeared to me to be excellent in that I noticed very ragged thruster firing on the first pass through all thrusters. I don't know why that was. The system should have been pressurized and we should have had solid fluid all the way out to all the thrusters, but, they were very ragged. The first trip around roll, pitch, and yaw, they steadied out to be very solid in firing. I'm sure that if they were ragged this shows on the data on the ground but I want to make sure somebody checks that. I'd hate to have that first portion of lift-off and not have very good thrusters during the very critical time of getting that baby off the descent stage.

11.0 CSM CIRCUMLUNAR OPERATIONS

11.1 Operation of Spacecraft

GORDON — About the only thing I can say on this particular one is that the operation of the spacecraft was excellent and I let the DAP fly the spacecraft practically all the time. Operation of the spacecraft by myself in lunar orbit was no particular problem or concern throughout the entire flight. The spacecraft systems were more or less taking care of themselves. There were no anomalies that caused any concern or required my attention other than an occasional glance to make sure that everything was functioning properly. I think it might be said at this time, during the entire flight after we had gotten on our way, we never had any caution or warning lights other than the O_2 high flow light.

11.3 Landmark Tracking

GORDON — The P22 for the first REV after landing has some excellent maps for the landing site for use during orbit operations. I was completely familiar with Snowman, the four or five craters that formed an arc downrange right in front of Snowman, so that when I came up for landmark tracking in P22, there were no problems in this regard at all. I had used the landmark 193 for the REV before DOI and first REV after landing. And that was just as easily done command module only as it was with the LM on.

The landmarker was easily recognized. There was no problem in tracking the landmark at all. The difference between the weight input with the LM docked as opposed to the ORB RATE torquing with the LM undocked, caused no significant

On board TV

difference in technique. The P22 on the next pass was to be at the landing site. Through most of the conversation, I wasn't really sure where it was. I had an update for the LM charts, giving the coordinates for what the ground at that time thought was the landing site of the LM, near Head crater. The targeting was pretty close to the actual spot where the LM had landed, but on the second pass after landing, when P22 came up, I found Snowman and I was actually looking at the Surveyor crater. Lo and behold, right there on the northwest edge of that thing was a bright shiny spot, a long shadow, and it was the only shadow in the area that I saw and as I got closer, it may be my imagination, but I thought I could see details of the descent stage and the landing gear extending from it.

As I approached overhead where the Surveyor crater was at the nadir, right in the center of that crater, and the dark shadow was one shiny bright spot that I knew had to be the Surveyor, this excited me quite a bit. I was pretty surprised that we were able to see that and I actually gave the coordinates on Surveyor back to the ground, which I thought was the LM landing site and it turned out to be exactly where they were. So once you know the general area, I should say pretty precisely the area in which they landed, anyone could find the LM itself in the sextant. Now there is a technique involved here. That is, first of all, not to search for the LM in the sextant. This is something I don't think can be accurately done because of the rate at which you're traveling over the surface and the field of view of that sextant. I had a good idea from their landing where they were. So, my technique was to find the area in the telescope. And when I found the Snowman on the telescope, I concentrated on the Surveyor crater itself, and positioned the telescope on the Surveyor crater; then transferred to the sextant. At that time, the alignment between the telescope and the sextant was outstanding. When I did go to the sextant, it was already pointed at the LM. The next pass around, realizing there would be a certain amount of skepticism about the ability to see the LM on the surface, I drug out the sextant bracket, put the DAC on the sextant and on the third time overhead, I tracked the landing site with the telescope, hoping to capture the LM and Surveyor with a 16-millimeter DAC (which I hope turned out).

11.4 MSFN

GORDON — The only thing I can say about this was, as expected, MSFN communications, transmissions, updates, and PAD messages to the command module were outstanding in all respects so there was never any doubt as to what was intended, wanted, or needed. In fact, I thought the flight itself, the conduct in this regard, was simpler than in the SIMS, with complete understanding on both parties as to what was going on and what was needed by each other. I have nothing but admiration for all those troops on the ground who handled this one.

11.5 Plane Change

GORDON — This was a plane change maneuver to establish the orbit for rendezvous day. I guess the thing to say about this was, once again, there was no particular problem associated with it. All worked perfectly. I realized at this time that it had been a real long day and I was tired and more prone to make mistakes. I certainly didn't want to be making mistakes during an SPS burn. When I came around this time and had AOS I chose to go to VOX operation and read the checklist, as I performed it, to the ground so they could monitor exactly where I was, exactly what I was doing, and would be abreast of the status of the spacecraft at all times. I previously had gone over about 5 or 6 minutes prior to the burn, according to the checklist, and turned on the bus ties and checked the LMP side of the space craft to make sure the functions over there were set in the proper position for the burn. There's no way I can monitor them during the burn but by reading the checklist over the air to MSFN. It gave me the assurance that I was reading the checklist correctly, not leaving anything out.

Now, I would think that the ground probably appreciated this. They knew exactly where I was in the checklist, what I was doing, and if I was behind and if I was ahead, so if any particular problem came up, they knew that I was with it or behind it. The 19-second burn was on time. Once again, the SPS engines performed like a dream, TVC DAP was shy. This was our first burn without the LM. The acceleration, of course, is much more noticeable than with the LM docked, but the guidance was excellent. The DAP action, gimbal drives, and the control of the whole thing were outstanding. There was no question in my mind at all that things were going as they should. In fact, there was no trouble monitoring that burn by myself at all.

Residuals were low. I don't remember what they were at this particular time. There was no trimming of the residuals, and I don't even remember if I copied them down. I'm going to mention the second plane change after rendezvous. That was the plane change for the bootstrap photography day. The only anomaly I noticed at all in any of the SPS burns was in this one and I'm not sure it's an anomaly. During the burn, even though guidance looked good and it was tight, it felt to me like the spacecraft was doing a dutch roll throughout the burn. It felt like a typical aircraft dutch roll-type thing. It was oscillating in roll and yaw.

CONRAD — I'm going to guess that we had some condition where the c.g. was passing through our low stability point.

GORDON — That was the only one. The next burn, the TEI burn, was as solid as a rock.

11.6 Update PAD and Alignments

GORDON — Now at this time, we were getting short on P30 PAD's. I think we had 12 P30 PAD's in there and we could have used twice that many. So Al had to manufacture some on the way back home so we could get all the P30 PAD's that were required. As far as alignments were concerned, the P52's were always done using PICAPAR in lunar orbit. It never failed to PICAPAR that I can recall and I always use those stars to align on. In fact, I think, Dnoces is one of the more common ones and it's probably the dimmest one up there and it was perfectly adequate to do the job. Once again, being in there alone, I didn't take the time, if the ground was watching the DSKY, to copy down any of the alignment PAD's and the flight plans as to the stars, star angle difference, or torquing angles. I made sure that the ground was copying the DSKY and they wrote them down on the ground, and I don't have to record them inflight. The torquing angles were very small and we were pretty happy with the performance of the IMU. Drift rates were very small and it was an outstanding platform.

11.7 Photography

GORDON — I guess there isn't a great deal to say about this. Most of the photography at that time was with the 70 millimeter and was concerned with targets of opportunity. There was no requirement to photograph anything specific during this time. The second day period was devoted to SO 158 and also to spectral photography experiments. That experiment was conducted with ease. The updates were well thought out and well planned. It was all conducted on GET, so there was never any doubt as to what exposures, what times the camera should be operated, and so forth. The ORB RATE torquing allowed me complete freedom and did not require any attention to the spacecraft as far as flying was concerned, once I had maneuvered the spacecraft to the right attitudes so that the hatch window was pointing at the nadir and used VERB 79 to start ORB RATE torquing. I virtually forgot about the spacecraft. The ORB RATE torquing is precise, and as far as I was concerned, it never got more than a half a degree off the proper attitude. It was an excellent control system to allow someone in there by himself to devote his entire attention to other things. I don't think this experiment could have been done without it.

That 158 equipment was easy to handle, easy to install, and easy to remove from the hatch window when a change of F-stop was required, and easy to put back in place. The target of opportunity S-158 experiment VERB 49 maneuvers were passed in adequate time for me to load the DAP, maneuver the spacecraft after the last series of S-158 was completed, and there was no problem at all in getting the so-called target of opportunity Theophilus, Descartes, and Fra Mauro with 158.

11.8 Monitoring Lunar Activity

GORDON — On AOS, Ed Gibson apprised me of what had taken place during the time I was out of communication, from LOS to AOS. He gave me a quick brief on what Pete and Al had done, where we were in a lunar time line, how things were going on the lunar surface, and when he caught me up to the present time, he let MSFN relay do its job. I must admit that whole communication system was outstanding during this time. The communications from Pete and Al were clear and excellent, and I didn't feel, shall I say, "left out" any time. It was well worth having this thought out ahead so I could know what was happening on the lunar surface, as far as the LM operations were concerned. I could talk to MSFN independently, without interrupting the LM operations. They weren't bothered by my communications with the ground. In fact, they didn't even hear them; at the same time it allowed me to listen to what they were doing on the surface.

11.9 COMMUNICATIONS

GORDON — The communications were excellent throughout this period. I might mention at this particular time, we had been having S-band problems before undocking. We started to pick-up some oscillation in the S-band antennas, the inability to maintain lock. It maintained lock, but it was oscillating in pitch and yaw so that the signal strength would decrease several decibels from peak. We did run tests on this later, on trans-Earth coast, but at that time we were operating the S-band in MANUAL so that every time I came around, we would acquire using a MANUAL mode and then go to AUTO and, being in BEAM WIDTH, I believe that in all cases throughout the pass, that S-band was never lost.

11.10 Maneuvering to Support Lift-Off

GORDON — One other thing that I wanted to do, just to see if it could be done, was to try and track the LM. If I had encountered it on the surface I wanted to track it from lift-off to insertion. The pass before lift-off, we were to do simultaneous P22's; subsequently we decided that we didn't have to because the LM did not do a P22 a REV before lift-off. I did try to take sightings on the LM during that pass. It was left up to me whether I wanted to track 193 or track the LM. I made the decision to go ahead and track the LM, since I knew where it was, and I knew I could see it, and could find it. I think this was a bad decision on my part, if we had needed that P22 information.

It turned out to be a bad decision because, in fact, I really didn't track the right landmark and I would have if it had been 193. The reason I didn't track the LM on this particular pass, I guess, was I got overconfident and forgot the procedures I had established before. I found the target first through the telescope and made sure, with the wider field of view, that I had the sextant on the right target. Well, I didn't do it this time. I had confidence in the state vector and the AUTO optics and P22. I relied on that to initially point the optics at the LM. Instead of going to the telescope and making sure that it was on the target, I went immediately to the sextant and unfortunately neither the LM nor the Snowman were in the sextant. I played with that a little while and couldn't find the LM and couldn't find the Snowman in the sextant. So, I went back to AUTO optics after trying to find it in MANUAL, looked in the telescope again, and by this time it was too late. I had gotten myself into a position where I couldn't find the right target. I had the Surveyor crater and when I went back to the sextant, there was a crater that looked like the Surveyor crater, but there was no LM and no Surveyor in the area. I took marks on this crater anyway, and I'm sure it was the wrong one, but I think I had time to take about three marks. That was a bad decision. I should have gone back to something I knew I could find easily. I should have tracked 193 and I recommend that space flights that need a P22 the REV before LM lift-off not to take a chance on finding that LM. Go back to the known landmark and do it the right way.

I was far too confident at this time and should not have done it. This is kind of unusual in this regard and that is, on the lift-off REV, I did it the right way again. I found the

Snowman in the telescope and put crosshairs on the Surveyor crater and actually at that particular time on the LM. So that when I went to the sextant, there it was. It was in the sextant. I didn't have to search in the sextant. It was already there and all I had to do was keep it in sight. What I was going to try was to keep the optics on the LM, reach up and go to FREE and then with the optics in MANUAL, so that the trunnion angle was approximately 22 degrees, I was going to try to use minimum impulse to actually maneuver the spacecraft to keep the LM in sight, but it didn't work. The ORB RATE motions were just fast enough at that time that I couldn't keep targets in the field of view of the sextant. I just flat lost it and once I lost it from the field of view of the sextant I couldn't find it again. So, I just gave that one up as a bad effort and immediately maneuvered to the insertion attitude. That's what I was doing from lift-off to insertion. When I got to insertion, I maneuvered to the P20 attitude which was approximately 118 degrees for the insertion attitude in pitch and then I maneuvered to 83 or 81 degrees for the P20 attitude, where I actually did the alignment. I looked at the alignment when I was at 118 degrees pitch to see if I could do an alignment there and I couldn't. The Sun/Earth combination of sextant to stars was not available, so I went on to the P20 attitude at 83 degrees pitch and the alignment was done there with no particular problem.

11.11 Rest and Eat Periods

GORDON — I guess, in this particular instance, the rest period was a relatively short one. For myself, it was started after the SPS plane change. To be perfectly frank, one guy in that command module taking care of all the activities that must be done, as far as your sleep period, it takes a considerable longer time than when there are three of you. I was scheduled for a 9½ hour rest period and I'm sure it was considerably shorter than that. That particular rest period was not inadequate, but it was certainly a short one. I was extremely tired that evening and could have used some more, although, it was certainly adequate to do the job. Our eat periods were in conjunction with this. It just takes time to prepare meals and eat them and then to clean up the dishes after you get through. We're all used to coming to tables, sitting down, eating, and then leaving and forgetting about the preparation and cleanup time involved. It does take considerable time. I did enjoy the rest period, however. I needed it.

12.0 LIFT-OFF, RENDEZVOUS, AIM DOCKING

12.1 LIFT-OFF

CONRAD — Lift-off went as advertised. I, probably more so than necessary, stuck my head in the cockpit and religiously punched OFF, every 30 seconds ENTER, and read out my parameters. We looked like we had a slightly hot trajectory. We were running a little high on V_1 and a little high on altitude all the way. However, that may be due to the fact that we were targeted for 37.0-ft/ sec R-dot to target for the zero CDH at rendezvous. Everything went absolutely as planned. At 200 ft/sec, Al went to open the shutoff valves and close the ascent feeds, and that's when I goofed up because he get a barber pole on the left main shutoff valve. I got interested in watching him instead of paying attention to my own checklist. I should have left him alone, and I didn't de-arm the ascent engine until we'd overburned 30 ft/sec. I knew exactly what I'd done, and I didn't wait for the ground I just backed it out. There was no reason for me to suspect that there was anything wrong with the PGNS because the PGNS and AGS were together all the way. They stayed right in there together.

BEAN — The ascent was pretty impressive. The only thing that bothers me a little bit is I had the camera mounted on the window bar and it was pointed out with pilot's eye view right at the horizon. I started it at about 30 seconds prior to lift-off, and then I noticed about 3 minutes into the flight that it wasn't running any longer. I don't know when it stopped, but I just hope it was running during the Pitchover, because during the pitchover you could look down and see the descent stage. You could see all the Kapton blowing all over the place. You could see the ALSEP that was still deployed down there.

It wasn't knocked over or affected a bit. It was a beautiful sight. I just hope it was running. I started the camera again, and it ran for 10 or 15 seconds and then quit. I started it again and it continued to run. I don't really know how much we have. It didn't look like very much of the film had been run out. I hope we don't miss that, but we brought the camera back with us.

CONRAD — That reminded me of something else. We had one anomaly at engine ignition. We had a MASTER ALARM with no light. I don't know what it was or what triggered it, and it had to be a transient event, but we did get the MASTER ALARM at lift-off.

BEAN — I don't remember it.

CONRAD — We got a MASTER ALARM. I looked up, there were no lights on, and I punched off the MASTER ALARM, and that was it.

BEAN — I don't think that I saw it.

12.4 Rendezvous Navigation

CONRAD — Okay. I don't really know that there's much to discuss in the rendezvous. We got inserted, and I backed off my overburn. The ground passed us a rough CSI solution or 46.5 ft/sec. We did an alignment, which was done just the same as the previous one. It was a good alignment. It was a four balls 1, small torquing angles. We called P20 and a first update gave a NOUN 49 that was very small. I incorporated it, and that's the last we ever saw a NOUN 49. We were right on the time line. We got into the radar at 36 minutes. We had plenty of updates for the CSI solution which came out to be very close to the ground-predicted solution. It came out very close to the command module's, and the rendezvous went exactly the way it's laid out in the checklist.

The burns were no problem. I burned them exactly the same as I did in the simulator. The simulator and the spacecraft were no different as I could see. I saw no differences in anything between the simulator and the actual flight all the way through TPI, and of course, we had absolutely no out-of-plane. We saw nothing greater than one fourth ft/sec throughout the rendezvous until TPI. A TPI solution came up with a 1.5 ft/sec, which I burned. As far as I can see, we were right down the pike all the way. The 1.5 ft/sec at TPI must have been a good one, because after the two midcourse corrections, which were both small, we made no line-of-sight corrections in either yaw or pitch. I just made none until we were a thousand feet from the command module. I never had to touch it. All I did was back off on the braking rates at 38 ft/sec. At 1 mile, I backed off to 30, hit AUTO braking rates on the way in, and from a thousand feet on in, I used my normals. I take 1 ft/sec off per hundred feet of range. We slid right on in there, and that was that. It was Mickey Mouse.

We had a tracking light failure. Al can give you PGNS and AGS residuals and burn parameters for CSI, CDH, TPI, and the midcourse correction. I'll talk about docking in a minute, and he can talk about his AGS updating.

12.5 LM and CSM Updates

GORDON — My only comment is that they were okay. I never had to have an update repeated to me specifically for the CSM. When all the updates were read to the LM, I just copied down those portions or those nouns that I needed for my own information. We had already made arrangements with Jerry Carr that when he read the updates to the LM he was familiar with those portions of the LM update that I required for the CSM, and he prefaced every one of those nouns by titling them, so that I would be alerted that this was in fact one of those that I needed. In other words, when he gave lift-off time, he just gave it as that. That was all I really needed, the lift-off time. He gave the whole lift-off PAD, and when he came to the CSI PAD, he merely stated it was the CSI time and the TPI time. All I needed out of those PADS was NOUN 11 and NOUN 37. The rest of the information was superfluous to the command module. It was required by the LM, and that way it was only read up one time. I picked out of the LM PAD's the information that I needed. This was done throughout, and it worked extremely well. I can't recall ever missing any of the PAD's that came up to the LM during this particular time.

12.6 Adequacy and Clarity of MSFN Data

GORDON — Adequacy and clarity of MSFN data was excellent and okay.

12.7 Updates for CSI

GORDON — Updates for CSI were good. I got the ground CSI solution with no particular problem. The problems that I had at this particular time were that I was actually getting a number of NOUN 49's, both from the VHF ranging and also from the sextant. I rejected the first one that came up and they were very small, although they didn't at the time meet the criteria of being less than 12,000 feet in 12 miles. When I looked through the sextant, I could see that I needed those updates. The LM appeared about half a radius from the center of the sextant and, in fact, I needed them so I got a number of NOUN 49's even after three or four updates or acceptances into the computer. I kept looking through the sextant, and AUTO optics was not pointing directly at the LM so I just accepted them. They were down in the relatively low numbers, and I guess that by this time they were actually less than 12,000 in 12, but I considered after the first few marks it was more or less the steady-state type operation and the criteria then was 2,000 in 2. I just made an arbitrary decision that they were needed and I took them.

Now, the first solution for CSI that I obtained was, I guess, a little out to lunch, and I guess it was a matter that we didn't get enough time yet because when I recycled at approximately 22 minutes, I was ahead in this time line. I was taking marks long before the 35 minutes, and this, of course, was allowed because the ground got the LM state vector up even before I had time to do the alignment. I was ready to do it and they had the state vector, so I took the state vector before I even did the alignment. The VHF broke out twice during this time before CSI, and I think that is the only time it did after the VERB 90 to get the out-of-plane which was virtually nothing. For the LM, I had plus 1.8 ft/sec and for myself, I had a minus 1.6. The out-of-plane distance was 0.32 nautical miles, so we weren't concerned about out-of plane at all.

My first solution for CSI was bad. It really hadn't converged yet. I got 38.8 ft/sec which I passed to the LM. This was calculated with nine VHF marks and 14 optics marks which I thought was plenty to get a fairly decent solution. It didn't, and I continued tracking, and I continued taking sextant marks. I just continued marking through the whole time period and I ended up with 14 VHF marks, 21 optics marks for my final CSI solution which finally converged and compared very favorably with the ground and the LM. I got a minus 45.9 ft/sec. The LM was 45.3 and the ground was 46.5, so we were all right there in the same old ballpark. It was interesting to note that after I had done the VERB 90, for the out-of-plane solution, that no longer did I get any NOUN 49's. I went right back and all the updates were acceptable and below the NOUN 49 threshold.

12.8 RCS/CSI Burn

GORDON — As far as I was concerned, everything was okay. The one hitch here was that the VHF communication was so bad during this particular period of time. For some reason, we did everything we could without having any effect on the communications at all. I never knew for sure that Pete was burning CSI. I naturally assumed that he was, but

I never had any confirmation and I kept asking him. I'm sure I bothered him by asking him whether he was making the burn or not, but I really felt that I ought to know. He could hear me, but I couldn't hear him. The only thing I could do, of course, was to assume that he had made the burn. Right after our CSI burn, we got communications back again. The P76 went in, AUTO optics was excellent, went right to the LM, and away we went tracking for the plane change.

12.10 RCS Plane Change

GORDON — Of course, none was required. I looked at the plane change after CSI. I obtained 12 VHF and 12 optics marks, did VERB 90's for both the CSM and LM, and got plus 0.4 ft/sec for the CSM, minus 0.4 for the LM and the out-of-plane was 0.2 nautical miles. Once again, no plane change was done. I continued marking until I had 16 VHF marks, 17 optics marks up to the nominal plane change time and then reinitialized the W-matrix, took a short series of marks and then recycled P33 at this time. For comparison purposes, I got plus 10.7, zip, and plus 8.3 for CSI, continued VHF and optics marks all the way up to the 10-minute time break before CDH, and with 17 VHF and 20 optics marks, once again the solutions compared extremely favorably. Command module solution was plus 10.3, 0 for Y, and plus 7 for Z. The LM solution was minus 10.2 and minus 9.3. So, everything was converged. State vectors looked excellent, AUTO optics was excellent, and we were on our way home free. There was no trouble monitoring the CDH burn, even though VHF communication was still bad. We were able to get through to each other after several tries. There was no particular problem after that, even though it was still pretty lousy communication.

12.13 Updating AGS with Rendezvous Radar Data

BEAN — The residuals on CSI was plus one-tenth, minus one-tenth. The burn was 45.3 for CSI, and the residuals were plus one tenth, minus one-tenth, and minus three-tenths. The AGS had a minus four-tenths, plus four-tenths, and plus six-tenths. The CSM solution there was 45.9, which converts to 44.9. That's pretty close to the 45.3. The CDH burn was a minus 10.2 DELTA V_X and a minus 9.3 Z, which compared with the CSM as a 10.3 and a 7.8. The residuals there were minus one-tenth, a zero, and minus two-tenths. The AGS saw zero, minus two tenths, and minus one. NOUN 81 was plus 25.9, minus 1.5, and minus 11.9. I didn't copy down the CSM solution here. While we're looking at the polar plot for the rendezvous, it looked to me like with a great number of points that, at the polar plot anyway, we were about 2 miles low. We stayed 2 miles low and slightly to the outside all the way in. It was just a nice neat rendezvous all the way. And we never crossed the line. We always just stayed about 2 miles or so and gradually got closer and closer as the rendezvous progressed.

The plan had been to align the AGS independently on the surface, keep it independent all the way through rendezvous, and then make a comparison and see how it did; not just to get data on the AGS, because the object is to rendezvous rather than get data on the AGS. So we did that on the lunar surface. It worked well at insertion. We did the PGNS alignment, but we did not align the AGS to it. It was in very good agreement with it anyway. I tried switching back and forth on my FDAI between PGNS and AGS, and I didn't notice any jump at all. I started taking rendezvous marks right on schedule.

The one concern that we'd had using the AGS was that I'd enter a RANGE for RANGE RATE or vice versa. So, in order to have a way out, we also recorded the information in the time line book. This would allow you to solve the CSI charts or CDH charts or TPI charts at the appropriate time. So, in case this did occur or the AGS did diverge, you'd be able to get a comparative solution. Well, sure enough, about the third mark, I entered either RANGE for RANGE RATE or RANGE RATE for RANGE and blew that AGS solution right apart. Fortunately, I'd copied down these values that I just discussed and I was able to go back and get a chart solution, which turned out to differ from the PGNS by only one-tenth ft/sec, which was pretty good. This showed that the charts were pretty accurate. But that right there shows the limitation of the AGS, in a system of this type.

CONRAD — You don't take that one unless you had a failure.

BEAN — If you take nine marks, it'd be 18 individual entries for CSI, and you can't stand one error. You got something that doesn't really do the job you want it to. After CSI, we realigned the AGS to the PGNS. Then I made all the AGS marks after that just as we'd planned to do, and got solutions that all compared very favorably. This shows that the AGS would do the job, would get solutions which we, of course, suspected anyhow. But the whole point is that you don't want to use the AGS as the normal rendezvous mode. It requires that every 2 or 3 minutes you make a lot of entries in the AGS. It requires that you point the spacecraft exactly at the command module, which takes time and effort. The LMP is working continually and isn't able to sit back and think through exactly what's going on in the rest of the spacecraft. It's way too much work. I think we need something better than charts that you solve as a backup manual system. We do need an automatic backup system. The system has to perform its work automatically, something like the PGNS does. Or, if it can't, it at least has to have the capability to perform a closed-loop solution without so many manual entries and so much work being done during the rendezvous.

I continued to work to input the data into the AGS until the second midcourse when Pete said, "Hey, why don't you quit working and sit back and enjoy the flight?" I got to thinking about it later and that was the first time I'd really looked out to see what was going on. The rest of the time I'd just been working my fanny off trying to get all those marks into the AGS, and that's not the way you want to fly a spacecraft. We did try to operate the AGS the way it was designed to do. We saw that, if you don't make an error, it does a good job. But it's just way too much work, and it takes too much time away from watching what's going on, just doing a lot of busy work.

CONRAD — The rendezvous radar low transmitter output didn't affect us. We got a lock-on as soon as we tried, which if I remember correctly, was about 235 miles. I'm sure the only reason we had the 17.5 DELTA-H was my screw up on the ascent shutdown because I didn't really trim out that 30 ft/sec too accurately. You know what happens to those residuals; they start growing and messing around. So, I backed off the 30 feet and got out of the program. I think the reason we didn't have a completely nominal 15-mile DELTA-H rendezvous was because of my screw up on the ascent shutdown. But the PGNS is quite capable in handling that, and it was no problem.

BEAN — Did you mention you didn't do a plane change?

CONRAD — Yes, no plane change.

BEAN — Did you say anything about the COMM?

CONRAD — No, we didn't cover the COMM. The VHF COMM for some reason became totally unsatisfactory. I guess we never did change configuration, did we? We always left it so Dick had VHF ranging, and something very definitely was wrong. Was my COMM clear to you, Dick?

GORDON — No, but it was readable.

CONRAD — Yes. You were almost unreadable to us. The other thing which I thought was very serious was, not only was the COMM garbled, but at CSI, Dick and I completely lost COMM with one another. He didn't have the vaguest idea whether I burned or not. I could hear him, but he couldn't hear me. I don't know what the problem was, but I think this VHF has to be looked at very carefully in both the command module and LM.

GORDON — I tried all sorts of different settings with the VHF. I tried different antennas and also tried different squelches. There wasn't a thing that seemed to enhance the COMM one bit.

12.14 CSM Monitor

GORDON — I guess the only thing I can say is it was easy; it was a fairly relaxed time. It was to me just about as simple as being back in the CSM. I conditioned myself to be that way, to exclude myself of external surroundings and confine myself to that lower equipment bay and occasionally to the MDC. As it turned out, I obtained a lot more than was actually required and could have relaxed a little on it.

12.15 RCS/TPI

GORDON — In the marking from CDH to TPI, there was some shafting in the telescope. I couldn't see anything; but with AUTO optics, when I went to the sextant, even though the field of view in the sextant was extremely white, after about 2 or 3 minutes I finally picked out a very dim white dot that was the LM.
I actually obtained eight optics marks before darkness. This was at the time that Sun shafting was supposed to preclude taking any optics marks, and I got eight between CDH and darkness. When we got to darkness, there were no more optics marks because the LM had lost its tracking light. I asked them to verify that it was on, and in fact, they had said it was. They cycled the switch with no apparent effect. The ground said it was taking power, but there was certainly no light. So from here on, it was VHF only. So with the eight optics marks, I obtained 16 VHF marks for the TPI solution. My TPI solution for comparison purposes was minus 26.0 in X, plus 1.7 in Y, and plus 11.1 in Z. I didn't write down what the LM solution was, but when I used their time option I came up with minus 26.1, plus 1.6, and plus 10.6 which is extremely favorable. It is within one-half ft/sec in the Z parameter and practically the same in the other two. It is interesting to note that the angle I came up with using their time option was 208.10 degrees as opposed to the nominal 208.30 for the command module transfer. It was extremely close. And the state vectors had still converged even though I had only eight optics marks.

12.17 CSM Sextant/VHF Track

GORDON — I guess the only thing I can say about that was it was kind of ragged. Once the state vectors got locked in, the NOUN 49's disappeared never to be seen again. Everything else was right down the track. The NOUN 49's only appeared for about a half a mark schedule prior to CSI; then they went away as I mentioned before.

12.18 Midcourse Corrections

GORDON — I let the P20 and P35 programs run with VHF only, and as expected, these parameters kind of blew up. Or the solutions, I guess I should say, blew up with VHF only, as I knew they would and as was predicted before flight with VHF only. My solution for the first midcourse was minus 1.6, plus 0.10, and minus 5.3. The LM's was minus 0.50, zero, and plus 2.0. So, you could see, it started deviating right there. The second midcourse, with VHF only, I had a solution of minus 6.1, plus 0.3, plus 1.6, whereas the LM was minus 0.9, minus 0.3, and minus 0.7. So, there's a data point; VHF only does blow up after TPI. But it gave me a chance to relax and just kind of watch the rendezvous from there because VHF, of course, functioned by itself and I merely monitored their solutions. Just to keep the preferred tracking axis at them so they'd have a full signal strength from the transponder, I just kept the P20 running until after the midcourse 2.

I did a VERB 77 P20 and allowed the DAP to point the X-axis of the spacecraft at the LM. By this time, of course, I think the state vector was off enough that it actually didn't point the X-axis at the LM, but there was no particular problem in climbing onto the couch and looking out the window and picking up the LM visually. I guess it was about 3½ miles.

12.22 Photography

GORDON — When I first picked up the LM visually, I had the television once again in the right-hand rendezvous window. I had the DAC in the left-hand rendezvous window, this time with the 75-mm lens, and I had about half a roll of color in the magazine. I exposed that at six frames per second until it was done and then I changed, took the camera down, and put 18-mm lens back on the DAC with a new, full, film magazine and let that run throughout the remainder of the rendezvous and throughout the docking. That was a slow-motion time. Pete had the rendezvous well under control. The braking was smooth and moderate, and there was no particular problem. I could see that the out-of-plane and all line-of-sight motions were locked. In fact, I took the COAS and put it on the LM at about 2 miles and I don't think he drifted more than a half a degree out of that COAS until he got right into station keeping where we were both starting to maneuver. As far as photography, television, and all those good things, there was no problem in doing all three of them - tracking, TV. As a matter of fact, I did most of the tracking with the TV monitor so that the folks back home could see this, rather than just me seeing it out through the COAS. I used the monitor to keep the LM in the field of view.

DAC

CONRAD — Did you take any 70's?

BEAN — Didn't have any cameras.

CONRAD — That's right. We didn't have any cameras. We threw them out.

12.23 RENDEZVOUS

GORDON — All I can say is that it was okay. Pete flew the whole thing up to station keeping distance; this was about 10 feet or so, 10 to 15 feet. I had already had a 60-degree roll in so that high gain antenna could be obtained for the television. When he actually stopped and started station keeping, I completed that roll to 180 degrees; and when I got there, I took over the active station keeping while Pete pitched over and did the yaw maneuver to essentially line up the docking target.

12.24 DOCKING

GORDON — I guess all I can say about the docking is it was as Pete and I both expected all these many months. In spite of all the arguments we got from, I guess you might call them contact dynamicists, about the dynamics of two vehicles coming together, one being 5000 pounds and the other roughly 35,000, it was a simple, easy task to perform. It could have been done in darkness as well as daylight with just as much ease, and the fact that the command module was active, rates were down low, the control was excellent. I used the DAP once again to dock with half a degree deadband and half a degree rate, and once I got the target lined up, my attitude corresponded to the LM attitude, and it was merely a matter of leaving my hand off the DAP and just translating. I thought the docking itself was a very simple task and relatively easy; however, at contact I felt it hit. I knew it was going to hit the LM; I just looked up and got the barber poles on the probe indicator and flipped the switches to FREE and didn't notice any motion in either vehicle during this whole time. I just sat there and watched it and let her stabilize and do what it wanted to do and it didn't wander hardly at all. I think I made a couple of pitch-up pulses with the rotational hand controller just to make sure that the alignment was exact when I retracted the probe. I don't think that the vehicles moved hardly at all at contact. There was certainly no noticeable motion, anyway. Retraction was as expected. It was a long period of time; it was very slow and easy, and there was no doubt that those capture latches or the docking latches had unmated.

CONRAD — We came right in and stopped. I pitched over and did the yaw maneuver after Dick did his roll. We did the docking just the way we stated before, and everybody objected to it. But we insisted on doing it that way, and I'm convinced it's absolutely right. Dick came in and docked; I maintained attitude hold tight deadband. As soon as he got his top latches barber poled, we went to FREE. Neither spacecraft so much as moved a muscle, and we got a complete, good lock. He straightened out attitude with his translations thrusters and went to hard dock, and it pulled us right in there without either spacecraft deviating; bango, we had 12 latches.

12.25 Post-docking Checks and Pressurization

GORDON — Post-docking checks and pressurization was all according to the checklist. There's no particular problem in doing all that stuff. In fact, I was down there as soon as we docked. I was in FREE and I changed the DAP. I had the ascent stage only and went to WIDE deadband and LOW RATE, went to plus or minus a half a degree and two-tenths deg/sec and just let it sit there in docking attitude while I pressurized the tunnel, verified its integrity, and completed the pressurization.

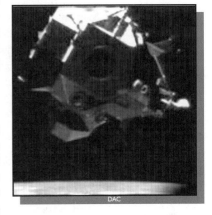

DAC

12.26 TUNNEL OPERATIONS

GORDON — I removed the hatch and stowed it under the couch. I reached up and had to take the pre load off so that the extended latch would engage. Of course, I knew what was going to happen, and at once it was engaged. It was merely a matter of waiting for Pete to get ready to take the probe and drogue into the LM. It came apart as advertised. There was no particular problem in this area at all. I might go back and mention one thing at this time. I had intended to wear my suit during the whole time I was in lunar orbit by myself. But I had gone so far as to take the struts off of the couch and put the EVA stabilizer bar in. I actually lowered the center couch. I lowered it to the floor. I couldn't stow it because of the SO-158 experiment or 810 stowage area. It had to be accessible the next day, so I knew I couldn't stow the couch at that time anyway, but I was going to lower it to get it out of the way.

The more I thought about this, the more I looked around, and the more I tried to thrash around with hoses on the suit and doing all those things by myself, the more I realized that it was absolutely ridiculous to spend all that time in lunar orbit wearing my suit. So I made the decision right then and there to take it off, and I did. I attached the lanyard to the zipper so if I had to put the suit back on again I could do it by myself. I knew that I could suit up by myself because the day of separation I had actually done so without any help at all and took less than 5 minutes.

Once I had the suit unstowed, the only thing I had to have help with was getting zipped. So I attached the lanyard when I took the suit off. I folded it up and stowed it in the L-shaped bag, and there it stayed forever. I thought about this EVA contingency thing and there was no reason to suspect at this time this was even going to have to be done. If it had to be done, I was just going to have to take the time to put the suit on and reconfigure the spacecraft and I figured it would take no more than 20 minutes total time. I felt that Pete could station keep at that time until I got ready. To be perfectly frank, it made the entire single-man lunar operation easier and more comfortable than expected. I hate to think of the shape I would have been in if I'd had to wear the suit all that time. And I had no qualms about docking or anything else with the suit off, because I'd already done it during the transposition and docking and the whole thing went as planned. In a nutshell, tunnel operations were okay. No hitches, no problems, and it went according to the checklist as planned.

CONRAD — Tunnel operations were smooth as glass. The LM was filthy dirty and it had so much dust and debris floating around in it that I took my helmet off and almost blinded myself. I immediately got my eyes full of junk, and I had to put my helmet back on. I told Al to leave his on. We left the helmets on and took off our gloves. Once we got stabilized and had the hatches open and everything, the flow system of having the command module more positive than the LM seemed to work. We did not pick up much debris in the command module; very little, if any, that was floating in the LM. But, it stayed very good in the LM all the way through our checklist.

We tried to vacuum clean each other down, which was a complete farce. In the first place, the vacuum didn't knock anything off that was already on the suits. It didn't suck up anything, but we went through the exercise. It did clean the rock boxes, that much I'll say for it. I don't think it sucked up any of the dust, but it brushed the dirt off the boxes. We put them in their proper containers, and transferred them. Dick brought over the LiOH B-5 and 6. We stowed those and it took a long time to get all the gear transferred. Then Al and I, because we and the spacecraft were so dirty, stripped naked and transferred the suits up to Dick. He stowed them under the couch and let us come in dirty and pack our own suits to keep himself and the spacecraft as clean as possible.

We packed the two suits in the lower part of the L-shaped bag, and to my knowledge we had very little debris come across from the LM. However, something we found out later and not until we got back to the ship, was that the fine dust was on the suits and on almost all of the equipment that was contained inside the bags. The dust is so fine and in zero g it tended to float off the equipment and it must have permeated the whole command module. It floated out of those bags; it floated out of the contingency sample bag. This we could see any time we opened up (which we stopped doing right away) the LiOH container that had the contingency sample in it. The whole thing was just a cloud of fine dust floating around in there. You could actually see it just float out of the bag through the zipper; and you can forget those zippers. They don't hold anything in. When we got all the gear back here and opened it up back on the carrier, we found out that it had all cleaned itself. That was where all this dirt was coming from in the command module.

The dirt is so fine I don't think the LiOH filters were taking it all out. It would pump it in the ECS system and pump it back out the hoses. This was indicated by Dick's blue suit hose, which we had tied over the left-hand side and was blowing on panel 8 circuit breaker panel. That whole thing was just one big pile of dust that was collected on the circuit breakers. The only reason it's there is the ECS hose was blowing on it. It's got to have taken the dust in through the LiOH canisters and filters and everything and blown it back out the blue hose. So the system is not doing the cleaning, the dust is too fine.

<u>12.27</u>
<u>Transfer of LM</u>
<u>Equipment</u>
<u>and Film</u>

GORDON — I think Pete has probably already commented an this during the LM portion of this debriefing, and to me it was a period of "hustle, hustle." I believe we had a lot more gear to transfer back and forth than did the Apollo 11 crew. Pete mentioned the vacuuming, the futility of trying to vacuum the suits and this sort of thing. He and Al both undressed in the LM and passed the suits over where we put them under the couches for stowage at a later time. It was a continual hustle to get ready for the LM jettison. A period of time after we docked, I had the tunnel cleaned out, but they were still busy gathering stuff up in the LM, so I went back and did the VERB 49 maneuver.

12.29 Configure LM for Jettison

CONRAD — We configured for LM jettison, and we got the LM off on time. Dick maneuvered right away to LM jettison attitude while we were doing the LM equipment stowage and everything. It was just a question of closing out and getting ready for it. The tightest part of the time line is the fact that it takes about 20 to 25 minutes to vent the tunnel down. We just got the tunnel vented down in time. We finally got it over plus 4, showing that we had a good vented tunnel. We got the LM off and said goodbye to Intrepid.

BEAN — That is a time period that ought to be lengthened up. We just didn't have enough time to get everything stowed, get back over, and get the tunnel depressurized. We just barely made it. We hustled like the devil. Tying down the bags and putting all the equipment in the LM just took longer than we expected. I think we were well trained to do it; it just takes more time at zero g. I'd recommend opening up the amount of time if we're going to do that again. There just wasn't quite enough time to do that job.

CONRAD — Stowage in the command module went very well. We knew where all the gear was supposed to go, and we just threw it all in the bottom of the command module. We got Al out, closed the hatch, and got rid of the LM. Then, while Dick messed around with LM JET and all that business, Al and I went to work and started stowing the gear. That's the one place the ground started bugging us. During that time, Dick was busy and Al and I were naked. We didn't have any clothes on; we wanted to get cleaned up after we had stashed everything because all the gear we were stashing was dirty. So we didn't have any COMM on, and I guess somebody got a little excited on the ground. They started bugging us, and that was a bad time. That's the one time that I got a little snippy, or Dick did. We shouldn't have, but we were very, very busy and on a very tight time line. We got it all done; everybody collapsed into bed that night.

GORDON — We maneuvered into a holding in the jettison attitude the whole time we were transferring the gear back and forth. There wasn't any doubt about what we had to transfer. There was no confusion in this regard about what had to go where, it was just a matter of there was a lot of gear and a lot of things to be done and not very much time to get done in. In fact, by the time Al finally got out of the LM, after setting up the computer and making sure that the LM was set up for the deorbit burn, we just barely made getting the hatch back in and getting the tunnel completely vented before separation. Now, I realize we could have separated without having the tunnel completely vented, but I wanted to have the tunnel vented before we jettisoned the LM, and we did get it done, but it was a tight time line.

12.31 Maneuvering to LM Jettison Attitude

GORDON — Maneuvering to LM jettison attitudes I already mentioned; the DAP did it.

12.32 Equipment Stowage

GORDON — We had stowed the spacecraft after the LM was jettisoned. Pete and Al did most of that while I was concentrating on getting rid of the LM by tracking it through the sextant.

12.33 TUNNEL CLOSEOUT

GORDON — Tunnel closeout was no problem.

12.34 IVA Photography

GORDON — We didn't take any IVA pictures.

12.35 SEPARATION MANEUVER

GORDON — The separation maneuver was photographed. We had the camera in the right-hand window that was taking pictures of the maneuver. It was an out-of-plane jettison maneuver. There was another P41 maneuver to the separation attitude where we fired 1 ft/sec retrograde for the separation. There was no problem keeping track of the LM. I couldn't see it all of the time out the left-hand window. Because of the out-of-plane attitude that we were in, Al could see it out the right-hand window. There was never any concern about a mid-air collision.

13.0 LUNAR MODULE JETTISON THROUGH TEI

13.1 Lunar Module Jettison and Trajectory

GORDON — There was no problem with the LM Jettison. The checklist was adequate. We were able to track and watch the ascent stage of the LM through the sextant. I took some marks to try to update that LM state vector as best I could. We wanted to photograph the LM descent and, hopefully, get the impact on the lunar surface. I don't think this was done even though we attempted to do it; we didn't have an adequate LM state vector using VHF ranging and AUTO optics to do this. I tracked the fire through the deorbit maneuver with the optics and VHF. I put a P76 in after the deorbit maneuver. I found it in the sextant and I took more marks. Then I put the sextant DAC on and tried to photograph it. I was watching the AUTO optics through the telescope, and I don't think we were too successful in preserving that LM state vector so AUTO optics on the sextant could keep track of it. I don't have any confidence that this worked.

13.3 Orbital Navigation

GORDON — I assume this is the P22 that was done on two REV's. All I can say about P22 is that it was easy to do - there were no problems. The maps I had onboard were adequate, and the landmarks were easily recognized. Tracking these through the telescope with the DAC on the sextant was easy. The ground suggested that I was taking marks a little bit earlier than I should. I got the T_2 time and I was not waiting the full 40 seconds to take my first mark. I did this on the first couple of landmarks and I observed that the trunnion angle was not as great as that of the sextant. I was being shortchanged when I went past TCA. I could judge this by the third mark that was supposed to be at TCA. I knew the time line was such that I should be taking the third mark right at TCA when the shaft was going through the 90 or 270-degree position. From that point, two more marks were supposed to be taken. I would wait 25 seconds and take the fourth mark, and then it would start to get crowded because the trunnion angle was getting out there near the 35 to 40-degree point. I could see the edge of the field of view peeking up towards the target, and I would wait and take that last mark just as the field of view in the telescope was right on the verge of losing the target. In the sextant, it would have been able to carry on for a little bit further. There is some discrepancy here about how far beyond TCA the telescope can be used for tracking. Tracking targets was a relatively easy task. I might mention that P22 navigation is done in RESOLVE and medium speed on the optics drive.

13.4 High Gain Antenna Acquisition

GORDON — High gain antenna acquisition, other than the S-band problem that I mentioned previously, did go a little unstable. We ran those tests on the way back and the ground has all that information. There was a heating problem. When it got hot, it did become unstable. When it did, we would have to go to MANUAL and WIDE. We would manually acquire at preset angles or leave it in AUTO and MEDIUM WIDTH. It didn't cause us any problem, although we did have a problem with the system.

13.5 OMNI and S-band Communication

GORDON — The OMNI and S-band communications were excellent throughout the flight.

13.6 Strip Photography Configuration

GORDON — Strip photography was easy to accomplish. Once we established the proper spacecraft attitude, we had ORB RATE torquing take over so that nobody had to fly the spacecraft. It was going in ORB RATE with the X-axis pointed at the nadir and the sextant DAC on. I was in the LEB trying to control the shaft and trunnion within 0 degrees shaft and 45 degrees trunnion angle. I would let it wander a degree or two and then drive it back to 45 degrees. The correlation between NOUN 91 and starting time was as planned. The times are written down in the flight plan. I think they're the preplanned times. Al operated the Hasselblad in the right-hand window and used the intervalometer. I don't think that he had any problem with that operation. Because of some of the other problems we had, we only did one REV of the strip photography whereas two were planned. We did have trouble with one of the 70-mm magazines. We had to go back and redo some higher priority high-resolution photography, and the ground had elected to eliminate the last REV of the strip photography.

13.7 Targets-of-Opportunity Photography

GORDON — I would like to make two general comments about this. First, there is no such thing as targets-of-opportunity photography. Either you are going to do them or you are not going to do them. If they're going to do this type of thing then we ought to plan to do it. There obviously are areas that are of higher priority and of higher concern to the scientific community than to those people at NASA on the ground. They ought to take the time to sit down and say here is your film, here are the targets we want you to take, and here is how we want you to take it. Don't leave it up to the crew to take this, that, or the other thing. We had two different types of maps and charts. We had a chart and strip-photo maps of the orbit in which we were. Each crew chooses the one they feel is better. I used them both, and I found that the chart was a good one for generally establishing where you were, what you were coming to, and at what time you should be over a certain area. I found that the photographs were more useful in identifying targets and allowing you to pick out those craters or whatever they wanted to be photographed. We need better maps and better charts for those areas. We should not call these targets of opportunity.

CONRAD — I agree with all that.

BEAN — I think the targets of opportunity ought to be made part of the flight plan.

13.8 Television

GORDON — We had no problem with the command module television. Our biggest problem was trying to figure out where to put it and the monitor. We had enough time in flight to figure that out. When we took outside television, it was generally in the right hand rendezvous window, and when it was on the inside, we found the best place. The television is here to stay, and it ought to be used

to its fullest advantage. I think it adds a great deal to the people's understanding of what we're trying to do and what is being accomplished. It's an excellent tool.

13.9 UPDATES FOR TEI

GORDON — No comment here. We were right on the money.

13.10 Maneuvering to TEI ATT.

GORDON — No comment. It was a DAP maneuver.

13.11 Sextant Star Checks

GORDON — The sextant star checks were all okay. We never had a problem with any sextant star chart throughout the whole flight. It was good.

13.12 Preparation for TEI

GORDON — By this time we were all prepared for TEI. We were eagerly awaiting the time.

13.13 THE SPS/TEI BURN AND ECO

GORDON — The SPS TEI burn and engine cutoff were right on the money. SPS performance was outstanding. The TVC DAP was solid as a rock. If I remember correctly, Pete told me that I had 1 second overburn, which I thought was pretty reasonable.

CONRAD — It was the first overburn we had.

GORDON — Yes, I guess that engine changes its performance.

CONRAD — We started out with a hot-engine shutdown early. The ground may have taken that into account and changed a couple of things.

GORDON — That's right. The burn time was 211. TIG was on time, and residuals were down. We weren't concerned on manual or anything else. It was well done.

13.14 High-Resolution Photography

GORDON — I'll call this one the high-resolution photography. Lalande was the first target after the lunar-orbit plane-change number 2. I fouled this one up completely. I was on a crater that looked like Lalande I didn't use my head. I wasn't paying attention to the times that the ground gave us. These times were accurate and were the ones we should have been using. I was looking at the ground, and this crater looked just exactly like Lalande, but lo and behold it wasn't; it was Herschel. We got some high-resolution photography of the southern lip of crater Herschel. We picked up Lalande during the REV prior to TEI. This didn't perturb any of the requirements for the rest of the bootstrap photography. I thought I should mention that. The pointing angles, the VERB 49 maneuvers, and the T_1-T_2 times that the ground supplied for the high-resolution photography on Descartes and Fra Mauro were excellent. They were right on the money. When I went to that attitude and a T_1 time came up, I was sighting out the COAS and it was right smack dab in the general area of what they wanted. There was hardly any maneuvering to do at all.

I did all the tracking in SCS minimum impulse and there wasn't any problem. It wasn't any problem at all to keep the COAS pointed at the target. You could pick out a crater on the ground, and the only time it became apparently difficult was when you were approaching the nadir or approaching TCA. With those high Sun-angles, the surface brightness was such that it was extremely difficult to maintain recognition of the crater that you were trying to track. There was a tendency for the surface to become generally washed out, but by judicious use of eyelids, eyeballs, and squinting, you were able to maintain a close resemblance to what you thought the target ought to be.

BEAN — While we were taking the high-resolution photography - I think it was after Fra Mauro - I had the 250-mm lens on. I had replaced the 500-mm lens with the 250-mm lens and was shooting some targets of opportunity. Between targets of opportunity, I had the camera in my hands and was rolling it about, and all of a sudden the side popped off the magazine. When it did, I tried to clamp it shut quickly but was unable to do so. I was unable to tell how much of the film was ruined by having been light struck from the opening in the side. We taped it up. I think that earlier, when taking some of the 50-mm photography, I may have operated the unlock mechanism instead of the film winding mechanism, which is on the other side. They both look the same, and I think this is a definite possibility for mistakes on future flights. I recommend that before any film is stowed onboard that a tape be placed over the lock/unlock mechanism. There's never any use in flight for unlocking the magazines. This will prevent any accidental opening or closing when trying to adjust the film winding. The same thing applies for lunar surface. There's no reason for the lock/unlock mechanism being exposed.

AS12-56C-8306

AS12-56C-8307

GORDON — I had a great deal of difficulty using the chart that had the targets of opportunity marked on it. I found it much easier to use the map that was made up of photographic strips. I really had a hard time using that other map - identifying where I was.

BEAN — I think the main problem was the difference in scale. It seems to me that you don't want to have two different scales up there. It's tough enough, in the short time you're there, to learn one. The thing to do is to select the scale map you would like. We all agreed that the scale 630 .000 to 1, which was the photographic-strip charts, was the better. If you make all the charts to that scale, you don't have to keep switching back and forth trying to remember which size craters will show up as which size on your chart.

AS12-56C-8308

CONRAD — You should carry two of them so you can continually pass it back and forth depending on which side you were working.

14.0 TRANSEARTH COAST

CONRAD — My only comment is that we should have come home in 2 days instead of 3 days.

GORDON — We had the fuel, too. We had 7 percent left.

CONRAD — Yes.

BEAN — This is when we were particularly glad we had that tape recorder onboard where we could listen to those tapes. I just wish that we'd taken more tapes, more batteries, and I wish somehow we'd taken a pocketbook to read, because there's a lot of loose time on your hands on your way home. You don't feel like debriefing and in our case, the Earth was about 1/16th full, so you couldn't see anything. The Moon: you'd been looking at it for 2 days and you didn't want to look at it again. You would have really liked to just rest and come sizzling home.

14.1 Systems

CONRAD — All transearth systems performed normally.

14.2 Navigation

CONRAD — Navigation again was no problem.

14.3 Passive Thermal Control

CONRAD — Passive thermal control was the same as on the way out.

14.4 Fuel Cell Purging

CONRAD — Fuel cell purging went as advertised.

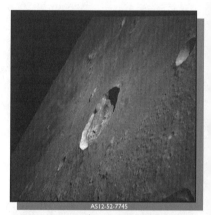
AS12-52-7745

14.5 Consumables

CONRAD — The consumable updates that we got from the ground were the same all the way throughout the flight, and they were satisfactory to us.

14.6 SPS MIDCOURSE CORRECTIONS

CONRAD — We made no SPS midcourse corrections.

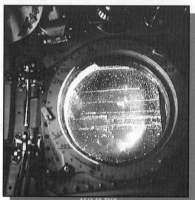
AS12-50-7369

14.7 Midcourse Lunar Landmarks

CONRAD — We did no midcourse lunar landmarks.

14.8 Star/Earth Horizons

GORDON — The P23's were done again coming back. The only thing to say about all that is that it's an eyestrain. I guess that's about it, really. It's well worked out.

CONRAD — Mention the fact that this was done primarily because of the proximity of the Sun to the Earth, and we never got those data before.

AS12-50-7372

GORDON — Those data come out in the wash, Pete. From the ground, I think. It was not hard to do. Some of the stars that I was given to look at were unobtainable simply because the light in the sextant, near the proximity of the Sun, was such that the star was not visible. These are identified and I think the ground, when we get the data back, will be able to resolve it so that we can see through the sextant, as far as proximity to the Sun is concerned. I might mention that the stars that they passed up with the unit vectors were no problem. You just use the planet option and stick in the unit vectors, and in all cases, if there were a star to be seen, as far as the lighting condition was concerned, it was always there. There was no problem with that at all.

14.9 DAP Loads

CONRAD — The DAP loads were as the flight plan published. Only every once in a while was there a difference. I don't think it even affected the DAP loads that much. I guess it did, once in a while, as the ground would call up a different set of quads to use, which is standard procedure for balancing, when we went into PTC or something.

14.10 IMU Realignments and Star Checks

CONRAD — IMU realignments and star checks were the same on the way back as they were on the way out.

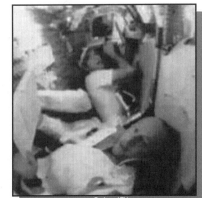
On board TV

14.11 COMMUNICATIONS

CONRAD — The COMM was good except for our little anomaly. We ran several S-band COMM tests by pointing the S-band antenna at the Sun for 4-hour periods of time to isolate our thermal problem which seemed to cause some trouble with auto lock and narrow scan.

14.12 Television

CONRAD — We did some television on the way back, which they had good reception on, as I understand it.

AS12-55-8220

14.13 Photography

CONRAD — We did some photography on the way back.

GORDON — I had in my written notes under the photography section, paragraph 13, all okay. I added that I wanted to talk at this time about this solar eclipse.

CONRAD — Yes. Well, I was going to cover this at the end of this thing because I think that's enough of a separate subject that we ought to do it.

GORDON — Okay. Fine.

CONRAD — The photography listed in the flight plan, or wherever it was, was carried out, and Al did most of that.

14.14 Eating & Rest Periods

CONRAD — Eating and rest periods were more than adequate.

14.15 Flight Plan Updates

CONRAD — The flight plan updates were nothing.

14.16 Maneuvering to Entry Attitude

CONRAD — Maneuvering to entry attitude was nothing.

14.17 Boresight & SXT Star Checks

CONRAD — Boresight and sextant star checks were done just like they're called out in the flight plan. You can do it in the simulator.

14.18 ELS Logic & Star Checks

CONRAD — The ELS logic was checked out at the proper time - so many hours before reentry.

14.19 EMS

CONRAD — We checked EMS. You ran two test patterns in one flight.

14.20 Entry Corridor Check

CONRAD — An entry corridor check was made.

14.21 Final Stowage

CONRAD — Final stowage was done. As a matter of fact, if you keep a clean spacecraft, you don't have much stowage to do anyhow.

14.22 Systems Verification

CONRAD — Systems verification was per the check list and flight plan.

14.23 Final Entry Preparations

CONRAD — Final entry preparations were per the check list and flight plan.

GORDON — I think you might comment that it's really an easy, relaxed, slow, methodical time line to follow. We had enough time in reentry to play chess. It's really well organized, and there's no particular problem with it at all that I can see.

14.24 SPS DEORBIT BURN

CONRAD — Number 24 is not pertinent because that's Earth orbit.

14.25 CM/SM SEPARATION

CONRAD — The CM/SM SEP went on time - 15 minutes before entry.

14.26 ENTRY INTERFACE

CONRAD — Entry interface was on time.

GORDON — Let's discuss the horizon check at the 267-degree pitch attitude, and the time, and so forth. It was dark and I never was certain that there was a horizon out there. I felt that there was one out there and that it really didn't make any difference. We had already checked the alignment. We were satisfied with the IMU. We had a boresight star. We had a sextant star check. We knew where we were, and the DAP was working properly. We were confident the whole time, and I didn't care whether I made that check or not.

CONRAD — We ought to change the rule, because we actually violated the rule.

GORDON — Well, we actually picked up the horizon check later on during the entry. That's that warm feeling once again - that you don't rely on anything and that you've got to look at the horizon to make sure you're pointed in the right direction. You don't do that when you're going to do LOI. You can't see the horizon on the Earth or the Moon, so there you go. I wasn't concerned about that, even though the horizon wasn't verified. We picked it up later in plenty of time before we got to entry interface where it became daylight. I was actually tracking the P22 or the NOUN 22 needles. The Moon was out there, and it went through the horizon. You've got the chart of the Earth horizon angles versus time from entry interface and you can check that any time prior to entry interface. It's a nothing check, and you can either do it or not do it. I could care less.

BEAN — Did you say we did see it later?

GORDON — I should say we saw it later. Oh yes. We didn't get it at that time, but we picked it up about 5 minutes later.

15.0 ENTRY

15.1 Entry Parameters

CONRAD — The reentry parameters were exactly as given to us. I think we hit our clock time right on the money, except drogue time. And that's got to be very hard to hit, because that depends on an absolutely nominal thing.

15.2 Communications Blackout

CONRAD — Communications blackout came on time. As a matter of fact, I gave them a call after we were supposed to have ended blackout and had Houston loud and clear.

15.3 Ionization

CONRAD — Ionization came when it was supposed to.

GORDON — We photographed that.

15.4 Control Mode(s)

CONRAD — All those were done according to the checklist. Dick flew it in MINIMUM IMPULSE.

GORDON — Well, maybe. I may have deviated here. I'll just make a comment. Once the service module was separated, I went to SCS MINIMUM IMPULSE and flew the command module back in-plane and tracked the NOUN 22 needles all the way down to 151 degrees in MINIMUM IMPULSE. Once we were satisfied everything was okay and the DAP was operating, I just went from MINIMUM IMPULSE to the DAP control for reentry.

CONRAD — Made Al Bean nervous because you never had gone back to RCS RATE COMMAND. But that's good that he was watching us.

15.5 Visual Sightings and Oscillations

CONRAD — We had a couple of fantastic visual sightings on reentry. Moonset for example; that really was spectacular. It's too bad we didn't have a camera to photograph that. It was a full Moon; and it was exactly aligned in the yaw plane behind us. Just watching that thing settle behind a beautiful, lit daylight horizon, with clouds above the Pacific, was phenomenal. Oscillations. I thought, boy, the spacecraft was steady as a rock. It's just like the simulator. I found that hard to believe and even made the comment I wanted to see whether the spacecraft was that steady and stable during reentry.

GORDON — It really is.

CONRAD — And it was just like the simulator. You get in the transonic region, and if you get it to oscillate a little bit, you get some more RCS firing.

15.6 DROGUE CHUTE DEPLOYMENT

CONRAD — Drogue chute deployment was quite late.

GORDON — We had 8:04 as the drogue deployment time; and I think it was 8:24, because I was counting it off to you and Al thought I was counting altitude, but I was counting clock time to get a hack on how late it really was.

CONRAD — They didn't hit that time very well.

15.7 MAIN CHUTE DEPLOYMENT

CONRAD — Main chute deployment was as advertised.

15.8 Communications and ECS

CONRAD — COMM was good and the ECS was good.

16.0 LANDING AND RECOVERY

CONRAD — We really hit flatter than a pancake, and it was a tremendous impact, much greater than anything I'd experienced in Gemini. The 16-mm camera, which was on the bracket - and we may have been remiss in this and I'm not sure, but it wasn't in the checklist - whistled off and clanked Al on the head to the tune of six stitches. It cold-cocked him, which is why we were in stable II. Although he doesn't realize it, he was out to lunch for about 5 seconds. Dick was hollering for him to punch in the breakers, and in the meantime, I'd seen this thing whistle off out of the corner of my eye and he was blankly staring at the instrument panel. I was convinced he was dead over there in the right seat, but he wasn't, and finally got the breakers in. By that time, we'd gone stable II which was no big deal.

69-H-1880

16.1 TOUCHDOWN - IMPACT

16.3 Postlanding Checklist

CONRAD — I went through the stable II postlanding checklist and up righted okay.

69-H-1881

16.4 Temperature and Humidity

CONRAD — I thought the temperature was extremely good in the spacecraft.

GORDON — Super.

CONRAD — Not like Gemini; Gemini really got hot in there.

GORDON — It really kept it comfortable.

CONRAD — We had the normal propellant smells in there from when we started sucking air back in the cabin on the chutes. We'd heard about that and didn't think anything about it.

69-H-1887

16.5 Communications

CONRAD — We had good reception on all frequencies. We had 64 hours of battery power.

16.6 Postlanding ECS System

CONRAD — We activated the postlanding ECS system per the checklist, and at that point, that was bad news. Well, I don't know whether it was bad news or not. We really didn't need to because the cabin was plenty cool and nobody was seasick, and nobody even got the slightest tendency to be seasick, although it was very rough out there. I would like to comment that I think that had I done this in the Gulf in that rough sea, I'd have gotten sick. I think that 10 days of zero g sort of numbs you to the motion sickness, and none of us even felt queasy. I could hardly get down in the LEB without feeling queasy in the Gulf egress exercise, and I was hopping all around that spacecraft out there in a sea three times as rough as it was in the Gulf. It didn't bother me at all.

GORDON — Do you think our vestibular response may have been desensitized?

CONRAD — Yes, I really do.

GORDON — For a period of time after zero g, maybe it's just not picking up these motions very well. We didn't take any medication or anything.

69-H-1893

CONRAD — No. No medication or anything. I could hardly get down the LEB. I had to get right back in the couch again in the Gulf.

GORDON — Let's go back to ventilation. The procedures say, of course, to open the PLV ducts. With that rough water out there, when we did, we just took water in through the intakes and that fan just blew it into the spacecraft. After a while, we got tired of getting wet so we just turned the PLV duct off. We just turned it off, and then when we got real warm again, I turned it back on just to let some more air in.

CONRAD — Until they got the collar on and that sort of stabilized things a little bit.

69-H-1599

GORDON — Yes. But we did take a considerable amount of water in through those ducts, even though we were in stable I position at the time before the collar was on. So, we just secured the postlanding vent.

16.8 Couch
Position

CONRAD — I think for the couch position everybody went to the 180, which was adequate. I eventually dropped mine to the 270 position once we got the gear in the spacecraft, and I think Dick did too. Al, you got dressed in the couch, right?

BEAN — Yes.

69-H-1604

GORDON — I got dressed back in the couch, too.

BEAN — Turned out to be no trouble at all.

CONRAD — We had plenty of time. We even got the medical kit out and put a Band-aid on Al so the whole world wouldn't get upset. As far as the inside of the spacecraft went, we were just waiting for the recovery people.

CONRAD — Apparently somebody keyed the primary 296 or whatever it was and we were receiving it, but nobody was reading us on primary except the Hornet. Now, I don't understand that.

GORDON — I don't either.

CONRAD — If somebody was keyed, I still don't understand why the Hornet was the only one that was reading us.

BEAN — We came up on primary A and B, so we had all our transmitters and receivers on.

CONRAD — Maybe the Hornet was reading us on secondary?

GORDON — The helicopter could read us.

SPEAKER — I'll comment on all that later. I don't want to put my comments on the tape, but I'll comment to you fellows.

CONRAD — Okay, but we had COMM, I mean we heard everybody. Apparently they couldn't read us, except on the Hornet.

SPEAKER — No. Apparently what happened was that Airboss, who was 25 miles to the south, was the only one that didn't receive you. All the HELO's heard you. The Hornet heard you on 296.8. His instructions are that if he doesn't hear you to tell you to switch over to secondary; and you switched over to secondary.

69-H-1888

GORDON — We had no anomalies in the spacecraft as far as COMM was concerned.

CONRAD — The COMM was super.

SPEAKER We feel the same way. We just played the recovery tapes back, and I'm as convinced as you are. No problem.

CONRAD — Okay. The decontamination procedures worked as advertised. The swimmer got the suits and the masks in to us. We had no trouble putting them on. It's one whale of an improvement over the BIG's, and I think caused us no problems whatsoever.

GORDON — You can see, and you're not hot.

CONRAD — I have a comment to make. I am really against inflating these life jackets. I went along with it, and Jerry Hammack, I guess, didn't believe us. He had the swimmers instructed to pull them on the way out the door, and that I object to strenuously. I don't want some swimmer pulling my life jacket when I'm halfway out the hatch, which he did, and it made me mad.

GORDON — Same thing to me.

CONRAD — Look, I know he was only doing what he was told to do, and that we've got to settle with the recovery people. There's no reason, when you've got a healthy crew in that spacecraft, to put them in that dinghy raft with those life jackets inflated, and it's bad news getting into that Billy Pugh net. I argued that with Jerry. I told him I'd go along with it, but I'm going to argue with them when I get back. I'm particularly hacked about having somebody pulling on me. Going out the hatch, a guy was hauling on me, and the next thing I knew, he inflated my life jacket. I don't want to say anything to them because they're only acting under orders.

16.12 BIOMED GARMENT OR COVERALLS AND MASK

CONRAD — As far as that mask goes, it didn't bother us at all. You can see well enough. We were, I think, in fairly rough waters. We didn't have any trouble getting our masks wet or anything because we didn't have those silly BIG's on, and we weren't stumbling all over ourselves unable to see where we were going.

69-H-1889

16.13 Spacecraft Power Down and Procedures

CONRAD — We powered the spacecraft down per the procedures, and the swimmer had a little trouble closing the hatch because the PLV vent cover got in the hatch. I think we should modify our checklist to remove that PLV cover completely and put it on the floor.

GORDON — Put it on the floor. That's a good idea.

16.14 EGRESS

CONRAD — The egress, otherwise, was okay.

16.15 CREW PICKUP

CONRAD — The crew pickup, I think, was excellent. The helicopter crew did an outstanding job considering the rough sea state. The swells were at least 15 feet, not 5, and the waves were a good 4 feet on top of that.

BEAN — I think one thing they should do is remove that sea anchor they have hanging down on that line about 10 feet below it, because it got in the raft. Then when they lifted up Dick Gordon, that line came whipping out of the raft, and I'd hate to have seen it if he'd gotten his arm in it or leg in it or something like that. It would have really injured us, so my recommendation is just to eliminate that completely. We sure didn't use it in our practice at all. We didn't need it.

69-H-1766

GORDON — That's right.

SPEAKER — Yes. That sea anchor bulkhead is supposed to be placed in the net after pickup raft, and the first guy jumped into it, about that time we'd slide on one of those 15-foot swells, and he'd be on his way like he was catapulted. It had nothing to do with the winch or the chopper.

17.0 GEOLOGY AND EXPERIMENTS

18.0 CM SYSTEMS OPERATIONS

The topics listed for this section are not discussed herein because they were covered chronologically in section 10 (Lunar Surface) of this volume. These topics will also be discussed in the follow-on photographic/ scientific debriefing that is published separately. Additional information about these topics can be found in the technical air-to-ground voice transcription, which contains all the EVA conversation as well as the informal inflight geological debriefing.

18.1 Guidance and Navigation

18.1.A ISS Modes

GORDON — ISS modes were okay; the only anomalies were during launch, and they have already been discussed.

18.1.B Optical Subsystems

GORDON — I think the optical subsystem worked well. There were no anomalies in its operation. I did notice a red spot in the telescope on the eyepiece. It was approximately 12 or 13 degrees up and approximately 5 degrees to the right. It was a little red chip mark. It was unattached; it didn't bother anything, but it was always there.

18.1.C Computer Subsystem

GORDON — The computer subsystem worked fine; no comments.

18.1.D G&N Controls and Displays

GORDON — The G&N control and displays gave us no anomalies, no problems.

18.1.E Procedural Data

GORDON — I thought our procedural data in onboard documentation per procedures, was excellent. We never lacked anything; in fact, there was far more information than we ever needed, but it was all well written and fairly concise.

18.1.F CMC/SPS/TVC

GORDON — The CMC/SPS/TVC was in all cases, except for one burn, good. It solved, and there was never any question about its capability to joggle this burn. Even during the LM burn, we noticed something in the lunar orbit plane change number 2; guidance was still good, however. A dutch-roll-type of oscillation was noticed throughout the burn. It appeared to be a classical aircraft-type dutch roll or a snaky oscillation, if you want to call it that in roll and yaw.

18.2 Stabilization and Control System

GORDON — On the subject of control, thrust vector control, and displays in loop control functions - these control functions all worked normally, without any malfunction.

18.3 Service Propulsion System

GORDON — The thrust switches were used on all burns. All burns were started on bank A, and then followed by bank B approximately 3 to 4 seconds after ignition. That system worked okay. The thrust vector alignment was always excellent in all starts. There weren't any trim transients noticed at all. The engine was trimmed from ground information for all burns. The DELTA V remaining counter and rocker switch all worked satisfactorily. The SPS THRUST DIRECT ON switch was not used; the DIRECT ON button was not used; and the THRUST ON button was not used. The SPS channel pressure indicator was good; there was no oscillation and the indicator worked fine. PUGS was good and operated satisfactorily. The only comment is to plan prior to the flight. The PUGS increase-decrease switch was operated in full increase throughout the mission.

18.4 Reaction Control System	GORDON — The service module RCS worked okay. No particular problems were noticed in the gauging side of the system. The propellant quantity gauge was too dark. It seemed to stay 100 percent throughout the flight. No problems were noticed in the command module RCS.
18.5 Electrical Power System **18.5.A Fuel Cells**	GORDON — The fuel cells have been better. The only anomaly we had was during launch when apparently a lightning strike knocked all three fuel cells off the wire, disconnected all three fuel cells from the bus. Their operation, however, was good throughout.
18.5.B Batteries	GORDON — The batteries were all okay; they accepted charges readily, and we had no problem with them. In fact, we practically never charged BAT C throughout the flight.
18.5.C Battery Charger	GORDON — The battery charger worked nominally; there was no problem with it.
18.5.D DC Monitor Group	GORDON — The DC monitor group was okay.
18.5.E AC Monitor Group	GORDON — The AC monitor group was okay.
18.5.F AC Inverters	GORDON — The AC inverters worked okay, that is, nominally throughout the flight. We never even used or checked out inverter number 3.
18.5.G Main Bus Tie Switches	GORDON — The main bus tie switches worked as advertised.
18.5.H Nonessential Bus Switch	GORDON — The nonessential bus switch was not used; in fact, all nonessential bus circuit breakers remained out throughout the flight.
18.5.I G&N Power Switch	GORDON — The G&N power switch was never used.
18.5.J Cryogenic System **18.6 Environmental Control System**	GORDON — In all respects, the cryogenic systems aboard the spacecraft worked nominally. There were no anomalies - there was an unbalance one day. The day Al and Pete were on the surface, I had to turn off heater number 2 on hydrogen and oxygen number 1. They ran for most of the day that way. They became balanced again, and I went back to AUTO. The system worked well; the ground did have to balance them a couple of times.
18.6.A Oxygen System and Cabin Pressure	GORDON — The oxygen system and cabin pressure worked well. There were no particular problems. The cabin seemed to stabilize at 5.1 Psi.
18.6.B Cabin Atmosphere	GORDON — The cabin atmosphere was okay. On the way out, it was clean. On the way back, we got lunar dust in the command module. The system actually couldn't handle it; the system never did filter out the dust, and the dust was continuously run through the system and throughout the spacecraft without being removed.

18.6.C Water Supply System

GORDON — The water supply system was okay, the hydrogen separator being on the spacecraft. We did use the gas cartridge separators on both the water gun and the food preparation valve. The one on the food preparation valve didn't seem to work too well, and it was removed during flight.

CONRAD — We think, though, that we might not have charged it properly.

GORDON — Well, you can't help but charge it. Every time you use it, you charge it. Water won't come out of there unless you fill it up.

CONRAD — Well, I'm talking about the technique we used.

GORDON — Yes, there is.

CONRAD — The best possible charge.

GORDON — We never really went back and recharged the thing, but I don't think any of us felt confident that it was really working. The one on the water gun was no particular problem to use. During the last two days of the flight, we removed both gas cartridge separators and never noticed any difference in the performance or couldn't tell any difference between their use or their non-use. I suspect it was because the hydrogen separator was performing satisfactorily in removing most of the gas from the water. We did get water or gas in the hot water all the time. However, this is not necessarily a problem. We did get a little gas in the hot water system, but maybe that was because we were heating water and forming bubbles in the heating system. The cold water didn't contain the gas bubbles that the hot water did. One of the big problems we had with the water supply system was the inability to shut water off readily at the food preparation station after use. That valve seemed to leak or drip water for 10 to 15 minutes after each use, and we were continually wiping it up with towels.

18.6.D Water-Glycol System

GORDON — On the water-glycol system, the mixing valve was put in manual, and the glycol evaporator temperature was placed at approximately 55 degrees manually through panel 382 on the way home because we were too cold. We used the manual valve to jack up the temperature a little bit. The water boilers were deactivated after TLI, and they were never operated again until just prior to entry. The crew never verified their function or operation, except that it was noted that the primary water boiler was operating. The steam pressure increased near 90 degrees K. Other than that particular fact, the crew never noticed the water boiler operation.

18.6.E Suit Circuit

GORDON — The suit circuit performed satisfactorily. There was no problem. The suit circuit was used during launch and undocking prep and was never used again. We did, of course, use the suit hoses as our ventilation system, and we put the screens over the exhaust hoses, which worked satisfactory. Going out, they were no particular problem. We cleaned the screens approximately once a day on the exhaust hoses and on the suit circuit return-valve screen. On the way back, with all lunar dust, this operation was increased to twice a day for the suit circuit screen and sometimes three and four times a day for the screens on the exhaust hoses. The screens were always covered with gray lint material that we never did seem to get rid of,

18.6.F Gauging System

GORDON — The gauging system worked properly. We had two failures in the gauging system and the environmental control system. The CO_2 sensor failed during launch and never worked again. The last day, the suit pressure gauge failed full scale low.

GORDON — There were no problems other than the pain of using bags to dispose of the feces. Other than that, the problems have always been the same with that system. We had the GFE URA system in addition to the old Gemini-type bags and also the launch day UCTA's which, of course, were used. At night, we used the Gemini system. Sometimes, when we didn't want to dump water during PTC, we actually stored it until the next morning when we did dump it. The GFE URA was used most of the time during waking hours and flight. It was very convenient to use and seemed like a worthwhile system. There is one problem, however. The URA always contained urine. The surface tension was such that even after you used it, it would have a film of urine on the inside of the cover; the Teflon top seemed to have a film of urine on it all the time and in general, even though it functioned properly, it was a sloppy thing to handle. Every time you used it, you came away with urine all over your hands. We were generally cleaning up with the towel and wiping off this excess urine that the system did not seem to do away with. The use of the GFE URA was easy and convenient. It was noticed, however, that the stream of urine had to be directed parallel to the honeycomb to prevent splashback. There were no cases where it actually backed up and overflowed. We did clog both of the filters of the urine system. Did the first one clog on the way out?

CONRAD — I think the first one went after four days or sometime when we got into lunar orbit.

GORDON — I thought it was way over half the mission.

BEAN — It was when we got back.

GORDON — That's right. It became clogged one day after Al and Pete got back from the lunar surface, and the second filter, which had not been used until then, was clogged one day from reentry. We used the system without the filter for the last day. Did you have a comment on waste management?

BEAN — I have a comment on that device. It would have been a lot better if it had been a little bit longer and a lot thinner. I noticed when we directed our stream of urine down the center honeycombs, the urine tended to go right through and dump. You could see the hose jump as the urine went out, whereas if you'd directed it down any of the other paths, I had the feeling that the urine was sticking around inside the device. I think making that change and the one that Dick mentioned might help. I don't know how you're going to fix it - perhaps some sort of band that keeps the urine that's down in the device from flowing up the sides of the device and then onto the cover where it sticks and makes it a little messy when you take it off.

CONRAD — Break the surface tension.

BEAN — That's right. You need a surface tension break right there so the urine doesn't tend to get on the cover.

GORDON — In case anybody's wondering about these clogged filters, we were fairly free and liberal in our use of fresh water to run through the system. It was, we felt, fairly well flushed and kept as clean as possible. We were continually purging the system. In fact, we would use it and leave the valve in the dump position and generally be reminded that we had done so because of the O_2 high flow light. We were continually getting those, and it didn't seem to bother the ground. They had commented on our oxygen usage, and it was a little higher than nominal. I think this is probably the reason for it. We had always purged that urine system quite freely and generally left it on until the O_2 high flow light came on.

**18.6.H
CO₂ Absorber**

GORDON — The CO_2 absorbers worked fine. Their usage, change, and account ability were all perfectly normal. We did forget to change cartridges one night, and the next morning when a cartridge change was scheduled, we changed both filters instead of one so that cartridge 15 had double duty on it and cartridge 17 only had half a cycle.

**18.7
Telecommunications**

**18.7.A
Monitoring**

GORDON — It was our desire as a crew to leave the monitoring and antenna selection up to MSFN for real-time control. It was our desire not to mess with antenna selections or antenna monitoring, except during certain phases of operations during the waking hours when we were to monitor and switch when we lost gain on any particular antenna. It really didn't cause any problems. There were a couple of cases where maneuvers affecting the antenna switching were not caught right away. There was a communication dropout a time or-two during the flight, but it never seemed to really cause any problems. We just felt freer by letting the ground take care of that. Al, do you want to say anything about management of the antennas or anything other than what I just mentioned?

BEAN — No. You've already covered the problem with high gain.

**18.7.B Individual
Audio Center
Controls**

GORDON — We never had any problem with the individual audio center controls. Everything worked okay.

**18.7.C
VHF**

GORDON — Our VHF worked properly throughout the flight, except during the rendezvous. We've mentioned in a previous section of the debriefing that during the rendezvous, reception was very poor. In fact, it made VHF almost unusable for communication between the LM and the command module during the rendezvous. We had to repeat, and sometimes the system was completely lost.

**18.7.D Operation
Of S-Band
High-Gain
Antenna**

GORDON — The S-band high-gain antenna was okay. There was never any problem. It seemed to operate satisfactorily for most modes. The anomaly in the high-gain antenna, which has already been mentioned, in the test we ran during translunar coast will help clarify some of those problems. It seemed that when we'd break lock, the antenna seemed to lose signal strength and would tend to hunt when it was in the REACQ narrow band. We had to go to manual for acquisition on occasion, and then it operated after acquiring in the AUTO medium mode. The ground has a handle on this because of the tests we ran during transearth coast.

**18.7.E Use of
CMC/DSKY To
Obtain Antenna
Pointing Angles**

GORDON — We did this most of the time, and when time permitted, we felt it was the best way to go rather than have the ground call up angles all the time. A VERB 64 can be used almost any time during most of the programs and unless we were particularly busy or something, we generally used that for high-gain pointing angles. However, when we were busy, the ground readily came up with pointing angles for reacquiring when needed.

CONRAD — My impression was that I spent about an hour one day in the simulator learning how to run that high-gain antenna. I think the simulator did a very good job of simulating how the high gain antenna actually ran in flight. I don't think anybody had any trouble with it in that the simulator high-gain operation was excellent. I think we were in good shape to run that equipment, thanks to good training.

18.7.F S-Band

GORDON — The S-band was okay. No problem other than those already mentioned was noted.

18.7.G Tape Recorders

GORDON — As far as this crew is concerned, tape recorders are probably one of the weakest areas in the equipment. I don't have a section for DSE operations. I'm going to use this section to talk about the Sony tape recorder we carried on board. Okay? We were very displeased with this. We never used a tape recorder to record any onboard data throughout the flight. We used the tape recorder for our own entertainment. We had several tapes with music that were made personally for us so that we could use them during the dull, boring hours during translunar coast and transearth coast. We used every single battery we had available in the spacecraft for this purpose. We had a lot of trouble with the tapes. Several times we had to stop the recorder and rewind the tape by hand. The batteries were good, but their lifetime is extremely short, and we felt this system, although not required, was not the best that could be utilized in the command module. As a crew, it is our recommendation that this is a very important matter, and steps should be taken to improve these tape recorders. It's suggested that a better tape recorder, not battery operated, be made available for this purpose. Instead, the recorder should use 28-volt dc power from the spacecraft.

18.7.H VOX Circuitry

GORDON — Whenever VOX circuitry was used throughout the flight, it seemed to operate satisfactorily. I did notice that the VOX setting had to be higher with the lightweight headset than with the COMM carrier, but this was no problem to adjust.

18.7.H DSE Operation

GORDON — The DSE operation was okay, and it was up to the ground to operate the DSE equipment most of the time, except when no particular down-link situation was being conducted in the spacecraft such as in SPS engine burn and entry where Al, in accordance with the checklist, used the DSE and select HIGH BIT RATE for recording purposes. The rest of the time, the ground controlled the operation of DSE.

18.8 Mechanical

18.8.A Tunnel

GORDON — No comments, really. It worked excellently, and the equipment functioned properly. There was never any problem with the tunnel equipment. The probe, drogue, and hatch all worked as they were supposed to.

CONRAD — I've got one comment, and I'm not sure you could improve on it. Everybody's aware of the water collection in the tunnel after LM jett. I got quite a shower bath, even though I had spent a great deal of time wiping up the water in the tunnel. I'm not sure that you ever get to all the water, but I thought that the Kleenex was not the most satisfactory thing in the world to wipe that tunnel down with, and I'm not sure that we shouldn't provide some kind of rag - I don't know if you can do this in zero g - and a bag that you can put the rag in and wring it out. We need something to mop out that tunnel with instead of winding up with a wet Kleenex. We had to wipe off the main hatch, and then the glycol panel next to the commander's couch, left-hand side, caught large amounts of water on it during the flight. I think we need to work up some means of being able to mop up water rather than just using a Kleenex that's onboard. I think we used at least a box and a half of Kleenex sponging up water.

18.8.B Struts

GORDON — The struts worked okay. We had no problem with them. They had been disconnected. The foot X-X struts were both disconnected from the couches on occasion. There was never any problem with disconnecting or re-engaging the struts. The locks on the main chutes were activated during entry, and there was never any problem with them. I guess that takes care of Command Module systems.

CONRAD — The inertial platform was excellent. We had low drift rates; we never had any problems with it. We had problems with the ACT. Alignments went well. The previously noted anomaly on the rendezvous radar, which was low transmitter power output, didn't present a problem. We weren't at maximum ranges, and that's the only noted anomaly on the rendezvous radar.

CONRAD — The landing radar performed in an outstanding manner. We had landing radar at 41,000 feet and it worked perfectly. We've already talked about the only noted difference between the simulator and the landing radar - that's the fact that the velocity rate went up shortly after the altitude light, which is not in accordance with the simulator. I'm sure the velocity updating didn't take place until 2000 feet V_I, but the light went out. This is a slight difference that ought to be cleaned up in the simulator.

CONRAD — The computer subsystem worked in an outstanding manner. We mentioned two uncalled-for alarms which seemed to be part of the program, and they showed up in the command module also. So, I'm sure it's part of the software. The G&N controls and displays all operated in a normal manner, and the only noted anomaly was in landing. The PGNS display of horizontal and lateral velocity did not work; and this time I'm not exactly sure why. Al and I are convinced that we were not out of configuration switch-wise. I don't know what the problem was there. As far as the procedural data went, the checklists were correct. The G&N dictionary (what limited use it had) was correct.

BEAN — The AGS operated exactly as we hoped it would. The only anomaly I noticed was that several times when we did an alignment with PGNS, we would check it with the FDAI, and it would be exactly the same. Other times, maybe it would be 10 minutes later, it would be maybe a quarter of a degree down, so you could pitch, roll, or yaw and the balls would jump. We used a confirmed alignment, right after that, with a VERB 49 NOUN 20, and they wouldn't seem to change anything much, so after a while if they didn't line up perfectly each time, I don't know. All the modes of operation seemed to work. I just discussed initialization. The AGS calibration and all the numbers were well within limits. We then performed part of descent with the AGS, twice, to see if we could get repeatable data, and sure enough, the data were right with it the second time.

BEAN — The problem of the AGS is that it takes too much manual work on somebody's part, mainly the LMP's, to close the navigation solutions. We did demonstrate, though, that if you do this and take the time to input the range and range rate at either 2 or 3-minute intervals, you can get a good solution. But I don't think this is the way you want to have a backup system in a spacecraft because it takes one man's complete attention to do the job. There are better ways to do it. One is the charts; I don't think the charts are what we want for a later system. We certainly want some sort of an automatic mechanical system that is better than what we have onboard now. I recommend using the charts, taking discrete data points, and then installing the charts manually to operate the AGS in the mode that they recommend. We gave the engine/no-engine commands to the AGS. We always operated on the PGNS. The electronics, the burns program, the controls, and displays all worked exactly as advertised. We were able to monitor the DOI burn, and we also set up the AGS to monitor all the rendezvous burns. They seemed to perform properly each time. During ascent, I noticed that the AGS needles were in complete agreement with the PGNS, and I had the feeling if we had to use the AGS for ascent, it would perform properly. At that burnout, the residuals were about the same as the PGNS during burnout.

19.2 Propulsion

BEAN — The propulsion system operated as advertised. The simulator is an excellent training device. It seemed to me that the engine came up after start at about the same rate the simulator did, and it operated pretty much the same as the simulator.

19.2.A Descent

BEAN — The only thing that we noticed that was different was during descent. We actually got more RCS firings than the simulator gives you, but we were prepared for this because Neil had pointed it out. I don't recall the DOI burn ever firing an RCS thruster. It probably did, and we just didn't notice it. The rest of the operation of the descent engine was nominal. The venting of the descent engine was much like that in the simulator except that the venting took much longer than it did in the simulator. I don't know the exact time that it would be on the tapes because the ground gave us a call. I don't think it would hurt to modify the simulator so that it vents at about the same speed because it looks like that's what we're going to be doing now.

19.2.B Ascent

BEAN — The ascent performed magnificently. There is no sound associated with it. As you lift off, there is a large bang as you separate the ascent and the descent stages, and then you just move rather rapidly as the ascent engine burns. It makes no noise. The ascent stage does a Dutch roll as it climbs out. It is not objectionable, and it actually feels very good as you are ascending. The RCS firing seems to be at about the same rate that you see in the simulator. There is quite a bit at first, then you pass through a period where there isn't any as the c.g. passes near the thrust vector; as it goes back out the other side, you start seeing more and more ECS firing. I think waiting until 200 ft/sec to open the main shutoff valves and close the ascent interconnects is cutting it a little bit close. When you're up in that air, you're getting fairly light in that you are accelerating fairly rapidly, and if I had to do it again, I think I would recommend that we open the shutoff valves at about 300 ft/sec to go and close the ascent feeds. You're not going to use much more RCS. You have plenty of it anyway, and you wouldn't run into any problems such as getting a sticky interconnect or having something else occur. There just doesn't seem to be any reason to wait until 200 ft/sec to go. You have plenty of RCS.

CONRAD — I feel that I have trapped myself by getting interested in Al's problem; but, by the same token, the engine redundancy is such that if the engine has been burning that well for 7 minutes, I think you should get the ascent ENGINE ARM switch off in sufficient time really to monitor shutdown. It was not apparent to me in the simulator that the vehicle was accelerating as rapidly as it was; and, for that reason, I got involved; I did overburn. That is my fault, but I feel that anytime you are within 400 ft/sec to go, you are well in an RCS capability, arid inadvertent shutdown is not going to do anything to you. You could relight the engine for that matter. I have a suspicion that the ground can check to see that all those relays are closed and give you a GO. The ground can verify that you have normal arming through the normal system, which is something that we never thought of; you should get that ascent ENGINE ARM switch off along with the main shutoff valve in sufficient time to handle sticky barber pole indications or to psyche out a problem with the ascent feeds or main shutouts. You should have sufficient time to get that ENGINE ARM switch off and get in AUTO shutdown.

19.3 Reaction Control System

19.3.A Attitude Control Modes

BEAN — As far as the reaction control system went, the attitude control modes, we did no flying in AGS. I didn't feel that it was necessary. It wasn't going to buy us anything on the flight. I felt that the simulator training, flying the LLTV, and everything else gave me sufficient knowledge of the AGS system. Other than checking it out to verify that it worked, we never flew in AGS, and I don't see that that's necessary. I used the most efficient system that I had at my means, and I think that is the way you should go.

19.3.B Translation Control	BEAN — Translation control was excellent through making burns and trimming. It was adequate station keeping in the little bit that I did on the command module after rendezvous with the ascent stage. All control modes, ascent and descent stage, I thought, were excellent.
19.4 Electrical Power System	BEAN — The batteries performed just as advertised with the exception of battery 5, which didn't seem to want to pick up the load of the system engineers bus when we put it on the line just before PDI. One other point on the batteries - we went into the LM early, powered up, took a look at the erasable memory, and gave the dump to the ground so that we could see if there were any effects of the lightning strike. As a result, the batteries were a little bit low when we powered them up DOI day, but it didn't seem to affect us other than that we went to high taps almost immediately after switchover from the command module. The switchover was nominal and everything else about the electrical power system was nominal. The dc monitor operated properly as did the ac monitor.
19.4.A.4 Power Transfer CSM/LM/CSM	BEAN — The CSM/LM/CSM power transfer was completely nominal the three times that we did it.
19.4.A.5 Abort Stage Configuration	BEAN — The batteries operated properly. We didn't notice any transients when we dead-faced the descent stage. The six main buses operated properly. The amperage on them was just about what we saw in the simulator. We never knew for a minute that anything unusual was taking place on the main buses.
19.4.A.7 Dead-facing	BEAN — Dead-facing was completely nominal and the only thing we saw was the talkback going barber pole.
19.4.B Explosive Devices	BEAN — The explosive devices, except for one, operated properly. We checked out both systems - one with deploying the gear and the other with activating the RCS - and it looked like both ED systems were operating in the proper manner.
19.4.C Lighting **19.4.C.1 Interior**	BEAN — The internal lighting was satisfactory. The only difference we noticed was the previously reported fact that the switch on the upper hatch turns out the floodlights. When we closed the upper hatch, it must have been out of rig. When we closed the upper hatch, after exiting the first time, the interior lights did not go out. We eliminated this by pulling the circuit breaker on panel 16. With that exception, everything was nominal.
19.4.C.2 Exterior Lights	BEAN — The exterior lights did not operate as well as we had hoped. By that, I mean the track light operated properly when we undocked because it was checked by Dick in the command module, and seemed to operate properly during descent; that is, the part of descent during which we had it on. The lights operated partially during ascent and the command module came up, right after CDH, and asked us if we had our light on, and if we were pointing the wrong direction. We tried to verify that the light was on by looking out the window, but I don't believe you can do that. We did ask the people on the ground. They indicated that the power was nominal for the light being on. We cycled the light in the circuit breaker, and it showed that when we turned the light on, the indicator light went off. When we finally rendezvoused, they indicated that the light was off. We pulled the circuit breaker, I guess. For some reason, that light had failed right around CDH.

19.5 Environmental Control System

BEAN — Oxygen and cabin pressure operated nominally the whole mission. We didn't have any trouble during the several cabin repressurizations, depressurizations and repressurizations that we went through. The pressure came out very rapidly, and stabilized. The repressurization portion of the system was nominal in all respects. The one caution here is to be aware of it when you switch from REPRESS, AUTO, to CLOSE and back again. You're going to get a very loud bang, and you'll want to be sure to let the crewmen in the command module know when this is taking place.

19.5.A Oxygen and Cabin Pressure

19.5.B The Cabin Atmospheres

BEAN — Cabin atmosphere from activation planning was excellent. When we got back inside the first time, in one-sixth g, the atmosphere remained that way although we brought in quite a lot of dust. The same with the second time and the cabin jettison depressurization. Once we got into orbit in zero g, there was a lot of dust and dirt floating around the cabin, and we chose to remain in our suit loops as much as possible because of all this dirt, dust, and debris that was floating around. When we finally got back to the command module and docked with the CSM, we wanted to figure a way to keep this dust and dirt from filling the command module, but we weren't really sure how to do that.

I think procedures should be developed so that a positive flow of air is maintained from the command module to the LM, not necessarily to keep lunar bugs out of the command module, but to keep all this dust and dirt out of the command module. We were plagued by it when we finally did get back into the command module. Pete and I had to remove our hoses so that we could use them for vacuum cleaners. Incidentally, they didn't perform too well. There wasn't enough vacuum there. We had to remove our helmets from our suits, to keep our eyes from burning and our noses from inhaling these small particles floating around; we just left our helmets sitting on the tops of our heads. This isn't a very good configuration to be in, but we had no other alternatives at the time.

I think this is completely unsatisfactory, and there must be some way to clean up that cabin atmosphere so that you can work in a good, acceptable environment when you do get back to the command module. It's possible that you could get up and dock with the command module before you open the upper hatch, dump the cabin down to 3-1/2 psi, and hope it doesn't blow a lot of the dirt and debris out of there, and then slowly fill the cabin up in the command module and that will keep it filled. There ought to be some way to do this job.

19.5.C Water Supply

BEAN — Water supply in the LM was good, cold, and adequately supplied throughout the whole flight.

19.5.D Water Glycol

BEAN — It proved nominal the whole flight. - The ECS, the glycol temperatures remained normal throughout the flight.

19.5.E Suit Circuit

BEAN — The suit circuit was nominal.

CONRAD — This is probably the same problem that Neil and Buzz had. I continually got water in my suit down on the lunar surface. This was written up on the Apollo 11 anomaly reports as their not having their water separators in the A position. We double checked the water separators numerous times and it must have something to do with the length of the Commander's hose condensing water out in the system because Al never got it. But when I disconnected my hose several times we got 2 to 3 ounces of water out my hose. It came out in big bubbles, balls of water. And so there was no doubt that I was getting water in my suit, and I think there's some kind of design deficiency. We need another moisture trap or something in the system because it's very uncomfortable. I got enough water in the feet of my suit so that it wet my feet. I never could drain water out of the hose. I used the recommended procedures passed up from the ground. I tried to drain the hoses and could never get any water to drain out of them, but then after I ran them connected to my suit for a while, I started feeling water again.

During the night, I only used the hoses intermittently to dry out my suit from my normal sweat, as I mentioned earlier. The next day when I hooked up on the hoses preparatory to EVA and after the EVA at ascent I didn't notice any more water. Somewhere in there, we got the water out of the system and it stopped doing it. It was only the first day up until our sleep period on the lunar surface that I had any trouble with water. But I think that this very definitely is a design deficiency and not improper operation of the ECS system, as it was cast off in Apollo 11 and something should be done about it.

19.6
Telecommunications

BEAN — We had no difficulty in monitoring the telecommunications. The S-band angles and the signal strength provided a good index of what was going on as far as the ground COMM was concerned. The VHF was completely satisfactory during descent. During ascent, we've already covered. The operation of the S-band high gain antenna was nominal. The only thing that I noticed that perhaps should be reviewed and modified was when we were in the AGS CAL attitude just before we were going to an attitude where Dick could track the landmarks just before undocking for DOI. I had my antenna in AUTO TRACK and when the thrusters fired, it would make it move around a good bit and the opportunity existed there for the antenna to lock on to a side lobe (which it didn't do), but it's probable that you would want to go to OMNI there, put the high gain antenna in SLEW and find the optimum attitude to prevent straining the antenna drive during this period. I really didn't see any reason to be in high gain at that time.

The S-band was beautiful the whole flight, particularly on the lunar surface. It seemed as if Houston were right there on the edge of the mare with us. It was fantastic. The VHF was good in descent. We both covered the VHF during ascent; namely, that it did not work at all. It sounded as if our receiver were being overdriven, similar to having a transmitter right next to your antenna. It kind of garbled it and overdrives it. In the LM, we tried switching antennas; we tried adjusting the A and the B squelch at all possible settings and it never did have any effect at all. We didn't want to change COMM modes because we wanted VHF ranging for Dick and we just kind of muddled through. It was not acceptable and that's one of the main things of the flight that should be fixed.

The audio centers operated properly. We had no problem with them. The flight recorders worked okay. We operated per the flight plan and, at the end of the EVA's, we checked to see whether the ground estimated if there was any recording time left. There was 10 hours total, and we operated in a CONTINUOUS mode for descent and then VOX mode during EVA's, because the EVA's were rather long. We suspected that the recorder might run out of tape before we finished the rendezvous. Whether it did or not, we don't know. Because the EVA's are going to get a little longer, there's probably going to be a desire to operate the recorders in a different mode during descent and ascent; otherwise, you're always going to exceed the capability of the recorder.

20.0
MISCELLANEOUS SYSTEMS, FLIGHT EQUIPMENT, AND GFE

20.1 Cabin Lighting System and Controls

CONRAD — I presume this is for both spacecraft. We found the cabin lighting adequate in the LM, and as far as I was concerned, I found the lighting in the command module adequate. There were only one or two occasions that I remember that we had to use the flashlight for something down in the back end of the spacecraft. The lighting wasn't inadequate enough to warrant any changes, do you think?

GORDON — No, it was perfectly adequate. In fact, I liked the lighting in there, but I have one comment on its use. I noticed that none of us used the EL lighting by itself. I think the EL lighting is a little harsh and we always did subdue it a little bit. We always used the EL lighting, of course, for the MDC, but we also used the floodlights in conjunction with it. It tended to soften up the whole panel.

CONRAD — I want to make another comment about the EL lighting. When I set my watches a couple of times at night with all floodlights out, I noticed that the EL lighting is adequate to read all that it's illuminating. Unfortunately, it doesn't illuminate any of the switches, and I would hesitate to ever throw a switch in there at night with the EL lighting only. I always wanted the floodlights up enough so that I could physically see where the switches were.

20.3 Event Timers and Controls

CONRAD — About an hour or two before lift-off, we noticed the tuning fork of the main MDC mission timer in the command module was intermittent. Finally, we all concluded that something was wrong with the timer. Nobody felt that it was a big deal, that it was the timer and not the CTU; with that, we launched. The next thing we noticed about the timer was that we thought we'd broken it. There were two cracks, one in the right corner and one in the left corner. In searching our memory, we all remembered that there has been a history of the faces of these timers cracking. If I'm not mistaken, it was in the LM, though. I thought it was the LM mission timer that cracked, and as a matter of fact, ours cracked and was replaced in LM6. Anyway, these timers are all the same, I believe. We had no trouble with the mission timer in the LM, and the mission timer in the command module continually lost time at a fairly high rate. After we'd sleep 8 hours, the timer would be off over 5 minutes. We never set the timer on the MDC as I remember, and at the end of the flight, it was off 2 or 3 seconds. That clock worked fine. The event timer in the command module never had a problem. It worked in an outstanding manner. I've heard that the command module timers have been great before, and we never had any trouble with that one.

GORDON — What about after launch when it reset to zero after the first lightning strike?

CONRAD — No, it reset to 23 minutes and 26 seconds or something; how it got there, I don't know.

GORDON — Really? I thought it went to zero and stopped.

CONRAD — No. It went to 23 minutes and something, and I finally reset it to zero.

GORDON — Oh, okay.

CONRAD — It had some horrendous number, like 23, on it. The mission timer in the LM worked in an outstanding manner. We never had any trouble with it at all. We never reset it. It worked fine. It ran all the way through landing and all the time we were on the lunar surface. It was a good clock. The event timer in the LM worked in a fine manner. We never had any problems with it either. I think that's about it for clocks.

20.4 Crew Compartment Configuration

CONRAD — As far as the command module crew compartment configuration, I thought the stowage was outstanding. I can't really think of any other comments on that, can you? I also thought the stowage in the LM was excellent. All fittings worked correctly; the only thing I noticed was that Al used to do battle with the mounting of the 16-mm camera over the window, and I'm not sure if that just wasn't because it was hard to reach.

20.5 Mirrors

CONRAD — We never used the mirrors in the LM. In the command module, we used the mirrors for shaving, mainly.

20.6 Clothing and Related Equipment

CONRAD — I'd like to make several comments on the clothing and related equipment. First, I understand that people are negotiating for a single piece flying suit versus the one that we have right now, and I think that's a mistake. We found several different temperature modes, depending on whether we were in PTC or not. We spent much of the operation wearing only our two-piece pants, long underwear, and the outer jacket off. When you have to go to the John, it seemed a lot easier to get out of the two-piece operation. I feel that the two-piece suit is better, and I think that is the general feeling of all of us. The two-piece suit affords you the opportunity of changing your own temperature a little by being able to take off your jacket. We operated quite a bit with the jacket off and the pants on. The pants were handy because of the big pockets; we carried a lot of gear in those pockets. The zippers were great; we carried our dosimeters, our toothbrushes, our spoons, and this type of equipment in there. Now, the other thing is there's no doubt in my mind that we ought to do something about all this electrical mishmash. A continuous source of irritation to me was the hookup of the lightweight headset to the gigantic connector, the stiffness of the equipment, and the fact that it didn't bend in the right directions for a normal hookup, even when those electrical connectors were form-fitted.

GORDON — I think we need another set of inflight coveralls. The one set, after a few days, gets so clogged with urine and so dirty that you just hate to put the thing back on. We also need more towels. It would be advantageous to include a second set of inflight coveralls to go along with the second set of constant-wear garments. It is refreshing to put on that clean constant-wear garment. I think we all got into those after the LM rendezvous was completed and after we had a chance to clean up a little bit. All of us enjoyed bathing, and we certainly enjoyed shaving. We each shaved three or four times during the mission.

GORDON — It's kind of a highlight during the day when you're in a grubby environment and you can't get clean. In this regard, the inadequacy of towels on board the spacecraft was acutely noted by all crewmembers. In fact, we didn't even have enough towels for one day. We used the towels for several purposes; one purpose was to clean up the urine that was always in the couch areas, on the suits, and on the GFE urine dispenser. It is our recommendation that at least two towels per day be afforded each man. There's no particular problem getting this on board for stowage. We found that once a day we liked to strip down. We'd strip down completely and use the hot water with those towels that we did have on board. We'd completely sponge down and give ourselves a bath. I don't think enough can be said for this type of thing and for the way you feel. We wanted to shave and bathe daily, on a regular basis, but we simply didn't have the equipment on board to do it. There is no reason why this equipment can't be provided.

20.7 BIOMED Harness

CONRAD — I had a reaction to the blue jelly which I'd never had before. I think everybody had a little bit of a reaction, which is kind of unusual. I know Dick never reacted before, and I didn't. Maybe there was something wrong with that load of jelly, for all I know.

GORDON — I had a very slight reaction or itching on the left axillary sensor. It really didn't bother me.

20.8 Pressure Garments and Connecting Equipment

CONRAD — Our pressure suits were okay. As far as I know, all the connecting equipment was okay. I made one comment about my suit not fitting right, which was my own fault. I should have insisted on fitting it in a LCG.

20.9 Couches

CONRAD — The couches were good.

GORDON — Couches were no particular problem; the mechanisms all worked properly, functioned properly. The center couch was lowered, and the X-X struts were disconnected and connected again. The mechanical functions of the couches all worked satisfactorily.

20.10 Restraints

GORDON — All restraint systems worked as advertised; there was no particular problem with them. In the command module, they were never in the way.

20.11 Flight Data File

GORDON — The flight data file was complete, more than adequate. We all thought that the inflight data file for both vehicles was in excellent condition, with two exceptions. Those exceptions were the P30 PAD's and the lunar surface maps.

20.12 Inflight Tool Set

GORDON — The inflight tool set was never used. In fact, it wasn't even unstowed; I guess it was at one time. We used tool E all the time. It was the only tool that we really used. Al used a crescent wrench one time to pry the finger brackets for the window covers. Those things just didn't fit at all in those windows. They were a constant source of irritation every time we put them up or take them down. This has been a constant gripe for almost every flight we've had in the command module. Those things have been fitted on the ground; they've been fitted in altitude chambers, and when you get them in flight, they can hardly be put in place. There must be something that either changes their configuration, such as the foam padding behind them or whatever it is, so that we can barely get those covers on. The ones that were particularly bad were the covers for windows 1 and 5. Other than that, the inflight tool set was never used. It was kept in its stowage place throughout the flight.

20.13 Data Collection

GORDON — There was plenty of space to write things. We used the flight plan almost exclusively, as it was the best place to record data that we needed, besides the forms and charts that we had for that specific purpose.

20.14 Thermal Control of Spacecraft

GORDON — Thermal control seemed to be generally okay. We did notice at night that it would cool off, warm up, cool off, warm up, and so forth. I guess this was basically because of the manner in which we conducted PTC in that there was occasion where one side of the spacecraft would get cold and then warm. We did notice this very instant temperature change on our bodily comfort as well. It was no big deal, but it was noticeable. Our configuration for sleeping took care of this problem. We generally left our clothes on and used the sleeping bag. However, when we powered down the spacecraft, it did seem to get a little colder than normal. In conjunction with this, the moisture going out was no problem at all. The amount of moisture coming back with the LM off of the nose of the command module tunnel area, the hatches, and the side walls was considerable.

20.15 Camera Equipment

GORDON — In most respects, the only thing we can say about the camera equipment is that it was adequate and excellent. It seemed to operate properly. None of the camera equipment in the command module gave us any problems, with the exception of one magazine, magazine S, that came apart. It's our recommendation that, once magazines are loaded, they should be taped, or at least the locking mechanism on the magazines be taped, so that they cannot inadvertently be actuated during flight, possibly with the loss of some very expensive pictures. The flight plan had procedures for photographing the ionization layer and chute deployment with the 16-mm DAC in the right-hand rendezvous window. There's nothing wrong with this, of course. We left it up throughout the entry and landing, and this is where we got into a little trouble. It actually came off during impact and hit Al over the right eye. We had looked at it; questioned it; and without really thinking too deeply, we decided that if it could stand a reentry, it could probably stand the impact. It was our mistake; we should have removed it. If anybody ever has any intention of using that DAC for pictures during rendezvous or during high-g loads, reentry, or landing, they had best remove it before touchdown.

BEAN — The LM camera equipment, 16-mm specifically, did not operate like it was supposed to. We turned it on prior to descent and it worked all the way through descent; I noticed it was still running after landing. It worked all the time we used it for EVA pictures. Every time I got back in the LM after the EVA, the camera was still running and had been for 3 or 4 hours, which is the normal mode. The time that it did fail was during launch. I turned it on approximately 1½ minutes prior to lift-off. Then, I noticed approximately 3 minutes after lift-off that the camera had quit running and that the little ball on the side, indicating film usage, had hardly moved. So, I feel pretty sure that we didn't get near as many photographs of the lift-off as we wanted. I tried to start the camera again. It started, ran for approximately 10 seconds, and stopped again.

We brought the camera back for evaluation, but it doesn't take very much of a failure mode in one of those cameras to really blow some very, very good movies that you go a long way to get. I think that's just what happened on the lift-off. We're going to have to do something about those cameras so they work 100-percent of the time instead of almost 100-percent of the time. On the two 70-mm cameras, we've already discussed the fact that the handle on one of them broke - not actually the handle, but the nut that holds the handle to the camera pretty straight forward. The thing that worried me most about the cameras was that we were getting a lot of dust on them. I was afraid we were getting dust on the lens, and we had no means whatsoever to clean it off. I think it would be definitely desirable to have a whiskbroom on the MESA. We could use the whiskbroom to dust off the suits, and perhaps the back of the broom could have something so that you could use to dust off the lens of the camera.

I suspect that as missions get longer, we're going to get some pretty good dust coverings on the lens of the cameras. Such a dust covering is going to degrade the photographs unless we have some means of cleaning the camera lens off, which we did not have in this case.

21.0 VISUAL SIGHTINGS

21.1 Countdown

CONRAD — There were no visual sightings during countdown. We mentioned the lightning. We also mentioned the water during the countdown between the BPC and the spacecraft.

21.2 Powered Flight

CONRAD — During the powered flight, we mentioned the first lightning strike. Apparently we didn't see the second one.

21.3 Earth Orbit

CONRAD — In Earth orbit, nothing unusual, nothing in translunar. We want to talk about the solar eclipse and the fact that we all were caught with our pants down. We should have had good camera settings and film available for that because it was certainly a spectacular sight.

21.4 Translunar and Trans-Earth Flight

GORDON — I feel very strongly about this. I think that someone, the crew as much as anyone, really dropped the ball on this. We knew this was going to occur before flight and we mentioned it. The people who are interested in this type of thing, if there was any interest in it, were very remiss in not planning further in this particular event. To us, it was one of the most spectacular things we saw throughout the entire flight. I'm sure there's obviously some scientific value in this type of thing. However, the reaction in this regard was virtually nil. In conjunction with this the response of the people on the ground, at the time that we reported this, was extremely poor. The crew was left on their own entirely to come up with guesses on camera settings, films, and film speeds. Repeated inquiries to the ground took a considerable length of time before any information was gotten out of the ground at all as to what type of film, what exposure, and what time settings to use on the cameras. It was a very poorly handled phenomenon we all knew about before flight.

21.5 Lunar Orbit

CONRAD — I'm sure in the scientific debriefing, if there are specific questions we're well prepared to answer them. I don't think that we saw anything unusual in lunar orbit. One thing that I noticed, I mentioned over the air already, were the black rock slides and the very white craters only appearing in a specific area. That's about the only one I can think of.

BEAN — The only thing I saw was practically near the site area, and it was the only place on the Moon that I saw anything that looked like it, was a different color. This was over in the Sea of Vapors. There, many of the craters appeared to be black surface near the rims of the craters. Out in that Sea, in several places, it looked, from a distance, like a black field covered with sort of a white snow. This was the only place I saw anything other than the nominal basic, either light tans or grays or whites, or whatever color you want to call the Moon, depending on the Sun angle. The only obvious difference in colors was the Sea of Vapors and the slides that Pete pointed out.

21.6 ENTRY

CONRAD — I think we saw the normal sights on entry, except for the spectacular moonset. It was a full Moon. And we commented on that.

21.7 LANDING AND RECOVERY

CONRAD — We saw everything we expected to see on landing and recover.

22.0 PREMISSION PLANNING

CONRAD — The mission plan was followed.

22.1 Mission Plan

CONRAD — There was never any question about the mission plan.

22.2 Flight Plan

CONRAD — I'd like to compliment the flight planning people. We saw no mistakes in flight. Chuck Stough and all his people did an outstanding job on it.

GORDON — The flight plan is one of the better ones I've seen, as far as information in it for the crew. We asked for a lot of it ourselves and the flight planning people responded in a very unique and unusual manner. They were excellent to work with. They did an outstanding job in response to our inquiries and questions during the flight. I thought those people did an outstanding job for us on this flight.

22.3 Spacecraft Changes And Procedures Changes

CONRAD — There were no late spacecraft changes that we weren't aware of and didn't agree with. The only one that I can think of is on the TV camera, and that was a drastic mistake. We did do battle with our own people, and those of you that I did battle with know who they are, to get the TV camera out of hock and you wouldn't let us use it. I think that's partially responsible for our foul up. We never saw any drawings of that camera, or had the vaguest idea of what it was going to look like. Maybe we would have picked up this F-stop thing sooner. Let's be smarter on the next one.

22.4 Mission Rules

23.0 MISSION CONTROL

CONRAD — No comments on the mission rules. We had our day in court with everybody. We all concluded, to our mutual satisfaction, that the mission rules were all right. Everybody was treated fairly and I'm thinking in terms of the changes to the descent rules with which we worked.

23.1 GO/NO-GO'S

CONRAD — GO/NO GO'S all came as advertised. They were preplanned and given on time.

23.2 UPDATES

CONRAD — The same can be said of the updates.

23.3 Consumables

CONRAD — My only comment is that we got several LM RCS consumable updates on the lunar surface; these were really meaningless, because we weren't using the ECS system.

CONRAD — We didn't get SPS fuel updates because we didn't ask for them. The one quantity that we asked for was the predicted helium values after the first big burn, the LOI burn. You know exactly where you are supposed to stand, because these had been discussed in advance. The quantity was passed up when requested. I think it was real good. There wasn't any comments on DPS other than which fuel gauge was to be the most accurate one for descent. That's something we practiced in SIM's and it came out fine. They called low fuel, just as advertised. I think everything was fine on that.

23.4 Flight Plan Changes

CONRAD — There were no flight plan changes before flight that we weren't aware of and agreed with; very few changes were made.

23.5 Real-Time Changes

CONRAD — The real-time changes were very few in flight and were necessary.

GORDON — I have a comment on the flight plan changes. It's always been my observation that daily flight plan updates usually came to the crew in the form of questions and usually these things were nominally written and carried out in the flight plan. It was our agreement with the CAPCOMM's and we assured FAO that we would knock off these so-called chit-chat sessions in the morning. If it appeared in the flight plan and it was nominal, that's what we were going to use and we didn't want to hear this information repeated. This suggestion or request was carried out quite well by the ground and our flight plan updates were those that were absolutely necessary for the conduct of the flight. I guess that covers the real-time changes.

23.6 COMMUNICATIONS

GORDON — We talked about communications enough and most of the time they were excellent.

24.0 TRAINING

24.1 CMS

GORDON — In a general statement, it was excellent. I don't think we were wanting in any particular area to simulate any particular phase in the mission that we felt was necessary prior to flight that the people in the CMS didn't come up with. They are extremely cooperative. They are capable and talented instructors, and they are easy to work with. They responded to our requests in all cases. I think the ability of the CMS to support lunar missions of this type is excellent. As far as the visual is concerned, it is excellent. The star ball and the MEP are generally good.

I would suggest that there are two areas in the command module simulator that could be improved. One of these is the P22. Landmark tracking needs improvement, and my suggestion is that they take the maps that are provided to the crew well before flight of the specific landmarks that you are going to be working with and make slides of these landmarks. You can use these in the actual mission SIM's so that the P22's you are training with will be exactly those which you will see in lunar orbit. That area got straightened out pretty well throughout the training cycle; P23 is the only anomaly in the CMS that is constrained somewhat in time and to Earth radius. There is some slight anomaly in what you can see, as far as the horizon-star relationships are concerned. By and large, I guess it's pretty good.

The docking was okay. We insisted that storage and configuration be up to date early in the training cycle. Larry Thompson and his people did an outstanding job of keeping the crew station in what we would consider flight configuration. Availability of the CMS was excellent in all respects. There was more CMS time available than I needed, required, or wanted. The ability of those people to run integrated with the EMS was outstanding, and we never missed a single mission SIM throughout the training program. It was excellent in all regards.

CONRAD — I go along with the CMS training. The instructors all did an outstanding job.

24.2 LMS

CONRAD — The same goes for the LMS. The mission capability was complete. The visual adequacy of the L&A was outstanding, and as far as the MEP's went, forget it. They never worked, and they are not needed. In the docking end, although I didn't practice much docking in the LMS, it always seemed to work all right when you wanted to use it. The crew station was good and exactly like LM-6. They did a good job on the configuration. The availability was outstanding.

24.3 CMS/LMS INTEGRATED SIMULATION

CONRAD — The CMS-LMS integrated simulations, we never dropped a one. My hat's off to the people who moved the stuff in Houston. The SIM software went over to Building 45. I was worried about that. It came back on the line in good shape, and we never missed a SIM. The same for the simulator people.

24.4 SIMULATED NETWORK SIMULATIONS

CONRAD — The SIM NET SIM's, I can only say that my opinion is still the same. The more of them you fly, the better the mission; and we flew plenty.

24.5 DCPS

CONRAD — The DCPS is a good training device. Six to 8 hours, and that's all that's necessary. Dick and I used it normally together, so that we worked as a team.

24.6 LMPS

CONRAD — I had no occasion to use the LMPS, although Al did run AGS rendezvous, and it ran okay. Due to our previous training on the D mission, we didn't feel it necessary to run PGNS on the LMPS.

24.7 CMPS

CONRAD — That's the same feelings Dick and I have on running the CMPS.

24.8 Centrifuge

CONRAD — I think the centrifuge training was a good run and worth the day we spent in it.

24.9 TDS

CONRAD — As far as I am concerned, the TDS is a waste of time. There's no need for it anymore. I think that if you have to dock the LM yourself you're in serious trouble and it's not that big a task. Perhaps an hour or two is all that is needed in the TDS.

24.11 NR Evaluator and GAEC FMES

CONRAD — The North American evaluator and the GAEC FMES are okay for a day or two to look at burns. Dick looked at some rendezvous, some tracking, and some burn information early in the game. Al and I waited until late in the game during our software checkout, then went up and flew DOI, PDI, and some ascents on the FMES. I wanted to see if there was any difference in the software, the computer lights, or anything else versus the LMS. There were a few minor points that we picked no.

24.12 Egress Training

CONRAD — If I go through that rubber room one more time, I'm going to go out of my mind. You can forget that one. Once is enough. We have more time in the rubber room than we do in anything else. If you have a crew that hasn't been through egress training, you ought to run them through the tank once, and that's enough. Gulf egress is always a good training exercise just prior to the mission, and I don't disagree with that one. The mockup egress at the Cape would be a good exercise if we had decent COMM and it could be run correctly. The COMM is terrible. The whole thing needs to be updated, and somebody needs to pay some more attention to it.

24.13 Spacecraft Fire Training

CONRAD — Spacecraft fire training is worth it one time around. We had 1 hour, and that's plenty.

24.14 Planetarium

CONRAD — We didn't use it because of our D-mission training.

24.15 MIT

CONRAD — The MIT operation, if it is up to speed; their simulator star ball does the job just fine. We held the G&N briefings at the Cape and they were all right. Every once in a while, someone would get off on a long-winded trail that wasn't necessary, but I think they are improving all the time.

24.16 SYSTEMS BRIEFINGS

CONRAD — We held systems briefings to an absolute minimum. Again, due to our D-mission training, all we wanted were the differences between LM3/CSM-104 and 108/LM-6.

24.17 LUNAR SURFACE TRAINING

CONRAD — Now we get to lunar surface training. I trained on the one sixth-g pogo, and I trained one-sixth g in the KC135. I thought these were excellent training devices for operating rock boxes and all our equipment one time around. As far as one-sixth g goes in the WIF, forget it. I think that's a waste of time, and it doesn't do the job. I think the WIF's okay for your zero-g egress training and practice, and that's about all I'd do on the WIF.

BEAN — A couple of real good things to run through about twice and towards the end of the mission when you want to see what it's like is the mobile pogo which is a very good representation of walking around on the lunar surface. The mobile pogo lacks Z-axis freedom, and that ought to be fixed because it tends to either tow you or drag you, and that's unacceptable. Another drawback of the mobile pogo is it doesn't have any uneven ground, and it's pretty easy to move around on the flat concrete. Those two changes ought to be made. The one in the centrifuge is quite a bit better. In particular, the rig that you get into that supports the weight of the backpack and you at one-sixth g is better than the mobile pogo. Maybe mobile pogo ought to take a look and see if they can adapt it? The ground over there at the centrifuge ought to be changed to be more representative of the lunar surface. But in either case, they are good training and I think they let you learn your normal pace. I noticed that I got on a nominal walking pace very rapidly once I got on the Moon. I was surprised how representative this simulation was.

CONRAD — I can't say enough about the one-g walkthrough suited exercises. I feel that the crew, the suit technicians, and the lunar Surface operations group working through Joe Roberts and Ed Gibson put together an outstanding training program for us. They spent a lot of time suited, as we did, and it's a direct contribution to the success on the lunar surface. Field trips were a boom. I want to compliment Uel Clanton and his troops on well-organized field trips. I don't think that we wasted any time. I think they learned, and I think we learned because we insisted on using our normal lunar surface tools and not make them straight geology trips. I think there was a little bit of education on both parts. I also want to compliment the Flagstaff troops through Al Chedister and Thor Carlstrom for all the maps they put together for us late in the game. They really came through, and this was real fine. I think everybody that's going to do lunar surface operation needs an actual run on the PLSS in the SESL. I know you've cleaned up the 8-foot chamber business and you do all this in the SESL now, and I think that's a good refresher for anybody. It gives you a chance to exercise your gear in a vacuum under heat conditions, and I think it's well worth the exercise.

24.18 CONTINGENCY EVA TRAINING

CONRAD — Contingency EVA training, did not use the KC135. We did use the WIF, which was satisfactory, and we did one-g walkthroughs. We may not have shown too much of this on our training records, but remember, we already had gone through this complete exercise once in D-mission.

24.19 MOCKUPS AND STOWAGE TRAINING EQUIPMENT

CONRAD — I thought our mockups were in excellent shape. The support that we had at the Cape from Houston and the Cape on our LM mockup was outstanding for the lunar surface operations.

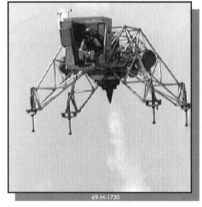

69-H-1730

24.20 PHOTOGRAPHY AND CAMERA TRAINING EQUIPMENT

CONRAD — We had cameras, and I feel we had the necessary equipment. We had enough time on that equipment to take it into the simulators and used it a little bit plus the lunar surface operations.

24.21 LUNAR SURFACE EXPERIMENT TRAINING

CONRAD — Lunar surface experiments training was satisfactory. The ALSEP training package did its job, and I think we learned a great deal from it.

24.22 LUNAR LANDING

CONRAD — I can't say enough for the LLTV. I really feel that that was the real frosting on the cake. It made me feel a lot better flying around up there on the Moon. I think the LMS does an outstanding job and where it breaks down in the last couple hundred feet, the LLTV fills in. It also gives you a chance to fly dynamically a vehicle that flies similar to the LM, I feel that the night flight up at Langley was worthwhile in the LLRF.

24.23 GENERAL SUPPORT

CONRAD — All our general support procedures, suits, checklists, and onboard data went well. I think everybody did his job. I also feel that the new procedures in handling the checklist, which we endeavored to implement and stick to, worked well. We had a few glitches in the beginning, but I think it's a very satisfactory procedure. I want to compliment the boys at the Cape that handled that data for us, the simulator fellows, Bob Pierson, Frank Hughes, and Glenn Parker, in particular.

24.24 PLANNING OF TRAINING AND TRAINING PROGRAM

CONRAD — I don't have anything to say about the planning and training; the flight speaks for itself.

BEAN — I think we ought to get regular TV to use for all the rest of the flights with a little monitor on it and use it inside and out, and I think by doing so we'll end up controlling the TV a lot better.

25.0 MEDICAL AND FOOD

25.1 PREFLIGHT

25.1.A Preventive Medical Procedures

25.1.B Medical Care

25.1.C Time for Exercise, Rest, and Sleep

25.1.D Medical Briefing

25.1.E Eating Habits and Amount of Food Consumption for F-5 + F-0

25.2 Food

25.2.A Appetite and Food Preference

25.2.B Food Preparation and Consumption

CONRAD — Yes. I'm afraid we let that one slip down the crack.

CONRAD — Preventive medicine procedures were, as far as we were concerned, fine when we needed them.

CONRAD — We really didn't need that much medical care, but when we needed it the doctors were there.

CONRAD — We had plenty of time for exercise, rest, and sleep.

CONRAD — The medical briefing was okay.

CONRAD — I don't think any of us changed our normal habits except we took the Dulcolax 3 days before the flight. Everybody went for the head the third day out on the flight and everybody but myself went two more times, right? Both you guys went 2 more times on the flight on about the 8th and 9th day, was it? I didn't, but it didn't cause me any discomfort and I think that was adequate.

CONRAD — As for hunger sensations in flight versus 2 weeks preflight, there was nothing. I didn't notice anything, did you? As for differences notable in food taste inflight versus preflight, I didn't notice anything there. I thought the food was excellent, and I'm sorry we didn't have the freezer and the cooker, but the food that was provided I thought was a whole order of magnitude better than Gemini food. My food preferences didn't change particularly in flight. The things that I didn't like before the flight I didn't like during the flight, and the things I liked before the flight I ate during the flight. That's just about the size of it. I ate a lot better than I did on my last 2 flights. I think Dick did also.

GORDON — Yes. I have one comment about the size of the food portions. I think Pete and I ate most of every meal. Al may not have eaten quite as much, but I would certainly think that the size of the food portions was more than ample. There was probably more food than is required or necessary. There is more food than we had been eating in the MQF and preflight as well. I think my preference, of course, was in the direction of one meal. It seemed to me that the spoonfuls where you could use hot water to prepare the food were probably the best type food. The wet packs, although good and tasty, would be much more appetizing if they could be warmed up. There's just something about chunks of meat immersed in cold gravy that is not nearly as appetizing as it would be if it could be warmed, although it was edible and tasty. I think all of us stuck real well to the menu during the first 5 days when we had total meals packaged for each individual. By the time we got around to the pantry, I'm afraid that we deviated somewhat from the preplanned menu in that it was somewhat difficult to look at the menu and then scurry around in the pantry for that particular item selected that day.

CONRAD — Now, wait a minute. Speak for yourself. I stuck very close to my menu.

GORDON — Well, that very well may be true. Al and I apparently didn't stick extremely close to the menu although we did try. I did at least try to find the food on the menu and if I didn't, it really didn't make any difference. I'd leave out one item and substitute other items for the menu. Al apparently deviated from his menu quite a bit, and Pete tried to stick extremely close to his in the pantry, but the first 5 days were very simple because the entire meal was packaged. I thought the pantry was a pretty good idea. It allowed us some latitude in the things we liked at the time. The amount of juices on board was certainly adequate. The fruits were extremely well done, and I think Rita Rapp and her people in CSD did an outstanding job of meeting our desires. Food on the whole was excellent. In fact, I don't think any of us hurt for food on this trip at all.

CONRAD — We noted some of the problems with dehydration, but I'll tell you where the problem came from. I don't think it was that the food wouldn't rehydrate. The problem was mainly with the hot food. When we got a little air in those packages, water saturation throughout the food was not good, and you tended to get some pockets. I thought food temperature was good on the hot food. I didn't notice any difference between the command module and the LM food. Did you, Al? We had wet packs; we also had spoon packages. The only thing that was missing was the hot water. I thought the spoonful packages worked extremely well. We had no trouble with food getting away from us, and I used the spoon - everybody used the spoon - all the time. I thought it was great. I used it in the LM and in the command module and I found it extremely handy.

25.2.C Food Waste Stowage

CONRAD — We had one waste food dump into the LM, and the rest of the food garbage went into B3. The last day of the flight, we put the 3 last meals, if I remember right, into the pantry. The germicidal tablet pouch worked well. We just kept that in one of our right-hand lower little under-the-eaves shelves and we'd pull out the package and pass the tablets around. I noticed no undesirable odors in the spacecraft. B3 was getting a little rank towards the end of the flight, but as long as we kept it shut it was okay. Now, the quantity of food eaten on the lunar surface varied according to the difference in individuals. I ate everything in sight, and Al didn't eat quite as much as I did. As a matter of fact, I ate a lot of Al's food.

BEAN — Yes.

CONRAD — I ate the whole ham paste tube and a couple of other things of Al's I really wanted to keep the old energy level up and I enjoyed the food; I wanted to eat it.

25.3 Water

CONRAD — We chlorinated at night and didn't drink after we chlorinated until the next day, and we never had a bit of chlorine taste in the water. I thought that was an excellent set of procedures. We've already griped about the leakage around the hardware, but that's another subject. We noticed no iodine taste, or at least, I didn't in the water in the LM. How about you, Al?

BEAN — The first drink.

CONRAD — First drink?

BEAN — Very first drink in the LM had a slight iodine taste. After that, it was cold and good the rest of the time.

CONRAD — Didn't notice any other taste in the CM or LM.

Physical Discomfort

CONRAD — We mentioned gas only in the hot water, I didn't suffer any physical discomfort from that sort of thing. I don't think anybody else did. I didn't notice any intensity of thirst during the mission, but we all drank heartily most of the time.

25.4 WORK/REST/SLEEP

BEAN — When I got ready to go to sleep at night, I never had a bit of trouble going to sleep at all. The only problem was that I'd wake up in about 5 hours and be wide awake. I had a choice then of either staying awake for 5 hours until everybody else got up or taking a sleeping pill. So I got in the habit of going to sleep for 5 hours; then I would wake up and take a sleeping pill and go to sleep for the next 5 hours.

CONRAD — We had more than enough time for sleep; there's no doubt about that. I got over on an 8-hour rest cycle and I never slept more than 8 hours that I remember, except maybe the first night, and I would just lie around the last 2 hours. I would sleep the first 8, and that was it.

GORDON — Sleep period programming was okay. Sleep is probably an individual preference. I definitely had a preference for actually sleeping in the couch. I slept in the couch all but two nights. These two nights, I slept in the sleeping bag underneath the number I couch, the left-hand couch. But it was always my preference to put the sleeping bag on, then get in the couch, and tie myself in the couch with a harness. For some reason, I slept better with the lap belt and the shoulder harness on, and securely lashed down to the couch, rather than free floating or being suspended in the sleep restraint under the couch. That was just a personal preference and it seemed to work better for me. During sleep periods, I would wake up maybe two or three times. I would look around the spacecraft and make sure everything was okay and then really go back to sleep.

The sleep period programming was more than adequate, particularly in translunar coast and transearth coast. I got extremely tired at the end of that first day of lunar orbit activities. That sleep period was scheduled to be a relatively short one anyway. It necessarily turned out to be so because at the end of the day was the SPS plane change, lunar orbit plane change number I occurred. But then I found that I had to do all the housekeeping and pre sleep activities by myself, whereas the 3 of us had been able to do them before and to clean them up in fairly rapid order. It took a considerable length of time to wade through all that by myself, and this cut short the sleep period. So I actually was pretty tired in lunar orbit and didn't really catch up until one day out of lunar orbit on the way back. I don't think anybody's performance was affected by fatigue and I'm not sure that fatigue really came into play. But certainly most of us in this particular occupation are used to performing while we are fatigued.

25.5 EXERCISE

GORDON — The thing we had for exercise, other than just moving around using the struts and the flat areas in the LEB for doing pushups and arm pulls or whatever you wanted to do, is the exergym. We all used it on the way out a couple of times a day for 15 or 20 minutes each session.

CONRAD — Al and I used it longer than that.

GORDON — Maybe a half hour each time, maybe a couple times a day. I didn't use it at all coming back. Al didn't use it coming back because the exergym rope was frayed. Pete was using it on the way back when he noticed that fibers were coming loose. So we elected not to use the exerciser at all on the way back.

CONRAD — What the exercise did for us on the way out was to prevent us from getting completely relaxed. Other than the discomfort in my shoulders which was the result of my suit being too tight, I never got stiff during any of the lunar surface operations other than my reported finger soreness. That was mainly from shoving them into the gloves for so long. Al and I took a look at our heart rates, and I would exercise until I was just getting warm. I didn't want to exercise heavily enough to really perspire up there because I wanted to keep my clothes as clear as possible. I think we exercised for periods longer than a half hour but at a slow, steady rate.

25.6 INFLIGHT ORAL HYGIENE

CONRAD — I guess everybody used his toothbrush to one degree or another. I didn't use it as much because my mouth doesn't get that bad in 100-percent oxygen. I did use the dental floss. I guess we all did. We all used the toothpaste.

BEAN — I liked the toothpaste.

CONRAD — I don't know where the rest of the guys kept their toothbrushes, but I just put mine back in my pocket after I cleaned it. I think everybody did.

25.7 Sunglasses

CONRAD — We transferred our sunglasses to our spacecraft uniform. The funny part about it is I used to use sunglasses all the time in Gemini. Orbiting the Earth and looking out at the ground, I'm used to the changes in color of the ground on the Earth, from flying and using sunglasses. I never used them in lunar orbit. I put them on a couple of times, but I didn't like the color that it made the Moon. I felt it degraded my observations of the lunar surface, so I never wore them. I don't think that any of us felt it was so bright that we needed them. The one time that I did put them on was as I took my helmet off at one mile from the command module. Anticipating it being very bright, I put the sunglasses on and put my helmet back on. I was sorry that I did that. The command module really wasn't that bright. That's the only time I used my sunglasses. I think I'm the only one who used them.

BEAN — Yes, you were. I kept mine off for the same reason.

CONRAD — That's an individual preference. I don't recommend putting them on or taking them off or anything. It's entirely up to the individual. I never noticed any trouble seeing instruments inside or outside the cockpit in any lighting conditions in daylight.

25.8 Unusual Visual Phenomena

CONRAD — I don't remember any unusual or unexpected visual phenomena or problems experienced that we haven't already mentioned. We all did see these corona discharges; and by paying a little attention to them, you could pin down that it was happening to one eye at a time. The discharges appeared in two manners. They appeared as either a bright round flash or a particle streaking rapidly across your eyeball in a long thin illuminated line. You either got a flash or a streak, and I could determine whether it was my left eye or my right eye that did it at the time. Most of the time I did this was during our sleep periods when we would be lying in our bunks. The next day we would either discuss it or write it up in the flight plan.

BEAN — One thing they wanted to know was how often and where. I didn't record where they were because it just seemed like anytime in the dark, if you wanted to, you could stay there a little while and one, two, or three of them would come by. If I was thinking about watching for them, I would see one every minute or somewhat less. One of them would be a flash, and about a minute later there would be a line. It didn't appear to make any difference whether we were in lunar orbit, translunar, transearth or anything else. If you just wanted to look for them, you could see them going by.

CONRAD — I do remember them the first night out, but I don't remember them the last night coming back. Otherwise, it didn't make any difference whether we were around the Moon or where we were. My only impression was that I noticed them more with my left eye than my right eye. Now, the other phenomenon which we didn't experience on VHF (and I'll be darned if I remember any VHF noises) was the whooshing sound that we had going down. We didn't have it coming up. It was a steady whooshing sound that was present on the front side of the Moon and the back side of the Moon. It didn't make any difference.

GORDON — The only time I heard it at all during the entire flight was just prior to LOI on the back side.

CONRAD — I don't think it was VHF. It was on S-band; and I think the other noise was on S-band, too.

BEAN — You don't have time to start troubleshooting those things and to try to figure out where they are.

25.9 MEDICAL KITS

CONRAD — We already talked to Dr. Jernigan about the aspirin in the medical kit. The idea they tried was a good idea, but it didn't work. And what you can do about that I don't know.

BEAN — There wasn't any aspirin in the LM where it really ought to be.

69-H-1763

CONRAD — Yes, I think you ought to put a bottle of aspirin in the LM medical kit.

25.10 HOUSEKEEPING

CONRAD — No new wheels to invent there. Housekeeping is its normal time consuming occupation which needs to be done. You can keep a neat spacecraft if you just spend enough time on it.

25.11 SHAVING

CONRAD — I think I shaved 4 times or 5 times. I enjoyed it and it made me feel better. I shaved twice on the way out and three times on the way back, if I remember correctly. I think everybody else did roughly the same thing.

GORDON — I think you guys shaved one more time than I did.

69-H-1765

25.12 Radiation Dosimetry

CONRAD — The personal radiation dosimeters were worn in our flight-suit pockets at all times. The readings were passed down for the record. We turned the radiation survey meter on once. We had a large increase like 30 millirads over the night before and, arriving on the water, all three showed it. So, we must have passed through something. We took the radiation survey meter off the wall one time to see if it worked, and it did. It didn't show much of anything. We put it back on the wall, and that's the last we ever paid any attention to it.

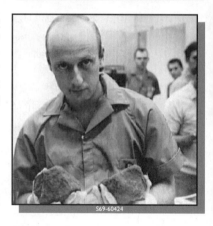

S69-60424

25.13 Personal Hygiene

CONRAD — I think the wipes should be bigger. I would personally prefer more towels and maybe a smidge less tissue.

GORDON — I would not decrease the amount of tissues. We used quite a bit of it; and if somebody gets a cold during flight, you're going to go through that tissue like it was going out of style. I don't think there is an oversupply of tissues in there. It was a general feeling that two towels per day is what ought to be allowed, and is required. Not only as a urine wipe when they are dirty, but also for bathing and cleaning.

CONRAD — It's more like you're going on the toilet. There are sure a lot of hang-ups. Hang-ups all over the place. The potable water was used for personal hygiene, and I'd also like to have some soap along for personal hygiene and just to get clean after lunar surface operation - just to get the dirt off. That's another reason we wanted more towels. We all stripped down all the way and washed down with the water and our towels several times during the flight.

BEAN — I have a comment on personal hygiene. The only thing I noticed on the whole flight that had any medical symptoms was the fact that when I got back up to the command module after the LM operation I had a pretty good rash. I washed it 2 or 3 days and it went away, but it surely did itch for about a half a day. I don't know what it was; I guess it was just the same as a baby's rash.

**26.0
MISCELLANEOUS**

CONRAD — I have no quarrels with the medical, PAO, or MQF requirements; they are satisfactory with us. PAO requirements didn't bother us particularly. I think the MQF operations have gone as well as can be expected to keep us cooped up for 5 days. I really don't think the MQF is built for a long-term stay. We have no comments yet regarding LRL operations.

APOLLO ACRONYMS*

AFD	Assistant Flight Director
AGS	Abort Guidance Subsystem
ALSEP	Apollo Lunar Surface Experiments Package
AOH	Apollo Operation Handbook
AOS	Acquisition Of Signal
AOT	Alignment Optical Telescope
APS	Auxiliary Propulsion System
ARS	Atmosphere Revitalization System
BCMP	Back-up Command Module Pilot
BIG	Biological Isolation Garment
BPC	Boost Protective Cover
CAL	Calibration
CAP COMM	Capsule Communicator at MCC
CCB	Change Control Board
CCFF	Crew Compartment Fit & Function
CDDT	Count Down Demonstration Test
CDH	Constant Delta Height
CDR	Commander
CDU	Coupling Data Unit
c.g.	Center of gravity
CM	Command Module
CMC	Command Module Computer
CMP	Command Module Pilot
CMPS	Command Module Pilot Simulator
CMS	Command Module Simulator
CO2	Carbon Dioxide
COAS	Crewman Optical Alignment Sight
COMM	Communications
CSI	Concentric Sequence Initiation
CSM	Command/Service Module
CTU	Central Timing Unit
CWG	Constant Wear Garment
DAC	Data Acquisition Camera
DAP	Digital Auto-pilot
DCPS	Data Collection Platform Simulator
DEDA	Data Entry and Display Assembly
DELTA-H	Height Difference
DELTA-M	Momentum Change
DELTA-P	Pressure Difference
DELTA-V	Velocity Change
DELTA-R	Change of Radius Vector
DELTA-VC	Change of Orbital Velocity
DOI	Descent Orbit Insertion
DPS	Descent Propulsion System
DSKY	Display & Keyboard
DTO	Detailed Test Objective
EASEP	Early Apollo Surface Experiment Package
ECO	Engine Cut off
ECS	Environmental Control System
ED	Explosive Device
EDS	Emergency Detection System
EECOM	Emergency Environmental & Consumables
EI	Entry Interface
ELS	Earth Landing System
EMS	Entry Monitor System
EMU	Extra-vehicular Mobility Unit (suit and backpack)
ETB	Equipment Transfer Bag
EVA	Extra-vehicular Activity
FDAI	Flight Director Attitude Indicator
FMES	Full Mission Engineering Simulator
FCOD	Flight Crew Operations Division
FOD	Flight Operations Division
G	Force exerted by gravity or by acceleration or deceleration
G&C	Guidance & Control
G&N	Guidance & Navigation
GAEC	Grumman Aircraft Engineering Corp.
GCA	Ground Controlled Approach
GDC	Gyro Display Coupler
GET	Ground Elapsed Time
GFE	Government Furnished Equipment
GSOPS	Guidance and Navigation System Operations Plans
H-Dot	Time Derivative of height (Altitude), Descent rate or Ascent Rate
ICDU	Inertial Coupling Data Unit
IGM	Interactive Guidance Mode
IMU	Inertial Measurement Unit
ISA	Interim Storage Assembly
Isp	Specific Impulse
ISS	Inertial Sensor System
IU	Instrument Unit
IVT	Intra-vehicular Transfer
KC-135	Training aircraft for low g simulations
L&A	Landing & Approach
LAM	Landing Area Map
LCC	Launch Control Center

LCG	Liquid Cooled Garment
LEB	Lower Equipment Bay
LEC	Lunar Equipment Conveyor
LET	Launch escape tower
LEVA	Lunar Extravehicular Visor Assembly
LGC	Lunar Module Guidance Computer
LiOH	Lithium Hydroxide air scrubber
LLRF	Lunar Landing Research Facility
LLTV	Lunar Landing Training Vehicle
LM	Lunar Module
LMP	Lunar Module Pilot
LMPS	Lunar Module Pilot Simulator
LMS	Lunar Module Simulator
LOI	Lunar Orbit Insertion
LOS	Loss of Signal
LPD	Landing Point Designator
LRL	Lunar Receiving Laboratory
LV	Low voltage
MCC	Mid Course Correction (or MCCH)
MCCH	Mission Control Center Houston
MDC	Main Display Console
MEP	Mission Equipment
MESA	Modular Equipment Stowage Assembly
MIT	Massachusetts Institute Of Technology
MPAD	Mission Planning & Analysis Division
MQF	Mobile Quarantine Facility
MSC	Manned Space flight Center
MSFN	Manned Space Flight Network
MTV	Manual Thrust Vector
ORDEAL	Orbital Rate Drive Electronics for Apollo & LM
O2	Oxygen
OPS	Oxygen Purge System
OX	Oxidizer
PAD	Pre-advisory Data
PAO	Public Affairs Office
PDI	Powered descent initiation
PGA	Pressure Garment Assembly
PGNS	Primary Guidance & Navigation Subsystem
PIA	Pre-Inspection Acceptance
PICAPAR	Abb. Pick a pair (of stars)Comp. routine
PIPA	Pulsed Integrating Pendulous Accelerometer
PLSS	Portable Life Support System
PLV	Post-landing Valve
PM	Phase modulation
POGO	Up and down oscillations/ POGO Suppression System
PSE	Passive Seismic Experiment
PTC	Passive Thermal Control
PUGS	Propellant Utilization & Gauging System
PU	Propellant Unit
PYRO	Pyrotechnic device
REACQ	Reacquire
REFSMMAT	Reference To Stable Member Matrix
RCS	Reaction Control System
RCU	Remote Control Unit
RDS	Rocketdyne Digital Simulator
R-Dot	Rate of change in range, rate of approach
REV	Revolution, one orbit
RLS	Radius or Reference Landing Site
S-IC	Saturn first stage
S-IIC	Saturn second stage
S-IVB	Saturn third stage
SCS	Stabilization & Control System
SESL	Space Environmental Simulation Lab
SEP	Separation
SIM	Simulator or Simulation
SM	Service Module
SPS	Service Propulsion System
SRC	Sample Return Container
STABLE I	Capsule floating nose-up
STABLE II	Capsule floating on its side
TCA	Thrust Chamber Assembly
TD & E	Transposition, Docking & Ejection
TDS	Docking Simulator
TEPHEM	Time of Ephemeris
TIG	Time of Ignition
TLI	Trans-lunar Injection
TM	Telemetry
TPI	Terminal Phase Initiation
TSB	Temporary Stowage Bag
TTCA	Thrust Translation Controller Assembly
TVC	Thrust Vector Control
UCD	Urine Collection Device
UCTA	Urine Collection and Transfer Assembly
URA	Urine Removal Assembly
VI	Inertial Velocity
WIF	Water Immersion Facility
Y-dot	Time derivative of pitch
W-Matrix	Position Error (ft), Velocity Error (fps), Radar Bias Angle Error (milliradians)

NOUNS

4	Gravity Error Angle
11	TIG CSI/T hrs:min:secs
17	Astronaut Total Att R,P,Y .01°
20	ICDU Angles
22	New Angles
33	Time Of Ignition
34	Time Of Event
37	TIG TPI
39	Delta Time Of Transfer Hrs, min, .01 sec
42	Apogee and Perigee
44	Apogee and Perigee Altitude
49*	Delta-R, Delta-V
65	Sampled LGC Time
69	Targeting Correction
76	Desired Horizontal, Radial Velocity Cross Range Distance
77	Delta-T to Engine Cut-Off
81	Delta VX,Y,Z, .1 fps
85	Velocity to be Gained (VG)X (up), VGY(Rt),VGZ(Fwd) in fps
86	Delta V in X,Y, Z

VERBS

1	Display Octal Data
6	Display Decimal Data
16	Monitor Decimal Data at half second intervals
32	Recycle
37	Change Program
49	Crew Defined Maneuver
63	Start RR/LR Self-test
64	Calculate Orbital Parameters
74	Initiate Erasable Dump via down-link
77	Set Rate Command/ Attitude Hold Mode in Digital Auto Pilot
82	Request Orbit Parameter Display
83	Request Rendezvous Parameter Display
87 & 88	Verb 87 and Verb 88 were used as a toggle on Apollo 11 to alternately lock out and release the VHF Ranging Data in and out of the computer
89	Start Rendezvous Final Attitude Maneuver
90	Request Rendezvous Out Of Plane Display
95	Inhibit State Vector Update (via Navigation)

AGC PROGRAMS

00	AGC Idling
12	LM Powered Ascent Guidance
20	Rendezvous Navigation
21	Ground Track Determination
22	Rendezvous Radar Lunar Surface Navigation
23	Star/Landmark Navigation Measurement
30	External Delta-V
33	SPS Minimum Impulse
34	Terminal/Transfer Phase Initiation
39	Proposed new program by Mike Collins
40	DPS Thrusting
41	RCS Thrusting
47	Thrust Monitor
51	IMU Orientation Determination
52	Platform Alignment
57	Lunar Surface Alignment
63	LM Descent Landing Maneuver Braking
64	LM Descent Landing Maneuver Approach
68	Landing Confirm
76	Target Delta-V

* Note: This acronym and AGC index was not part of the official documentation. It was compiled by the editor using a variety of sources, consequently there may be some errors. Kind assistance from Phill Parker, Frank Sietzen, Andy Chaikin, & Glen Swanson.